Practical R...

RELATED FOCAL PRESS TITLES

Elizabeth Czech-Beckerman *Managing Electronic Media*
Ken Dancyger *Broadcast Writing: Dramas, Comedies, and Documentaries*
Donna L. Halper *Full-Service Radio: Programming for the Community*
Donna L. Halper *Radio Music Directing*
William Hawes *Television Performing: News and Information*
Robert L. Hilliard *The Federal Communications Commission: A Primer*
John R. Hitchcock *Sportscasting*
Andrew F. Inglis *Satellite Technology: An Introduction*
Carla B. Johnston *Election Coverage: Blueprint for Broadcasters*
Michael C. Keith *Selling Radio Direct*
Marilyn J. Matelski *TV News Ethics*
Marilyn J. Matelski *Daytime Television Programming*
Robert B. Musburger *Electronic News Gathering: A Guide to ENG*
Ted E.F. Roberts *Practical Radio Promotions*

Practical Radio Promotions

Ted E.F. Roberts

Focal Press
Boston   London

Focal Press is an imprint of Butterworth-Heinemann.

Copyright © 1992 by Butterworth-Heinemann, a division of Reed Publishing (USA) Inc. All rights reserved.

No part of this publication may be reproduced, stored in a retrieval system, or transmitted, in any form or by any means, electronic, mechanical, photocopying, recording, or otherwise, without the prior written permission of the publisher.

 Recognizing the importance of preserving what has been written, it is the policy of Butterworth-Heinemann to have the books it publishes printed on acid-free paper, and we exert our best efforts to that end.

Library of Congress Cataloging-in-Publication Data
Roberts, Ted E.F.
   Practical radio promotions/ Ted E.F. Roberts.
     p. cm.
   Includes glossary
   ISBN 0-240-80090-7 (alk. paper)
   1. Radio advertising.   I. Title.   II. Series.
HF6146.R3R63   1992   91-37960
659.14′2′—dc20   CIP

British Library Cataloguing in Publication Data

Roberts, Ted E.F.
   Practical radio promotions.

   I. Title   II. Series
   659.142

   ISBN 0-240-80090-7

Butterworth-Heinemann
80 Montvale Avenue
Stoneham, MA 02180

10 9 8 7 6 5 4 3 2 1

Printed in the United States of America

Contents

Preface   vii

Introduction   ix

1 **The Growth of Radio Promotions**   1
   The Golden Age of Radio   1
   The Radio Networks Arrive   2
   Television Arrives   3
   Conclusions   6

2 **Promotion Management**   8
   The Management Team: The Promotion Director   8
   Promotion Director: Qualifications   11
   The Promotion Budget   11
   Conclusion   13

3 **Selling the Station**   15
   Developing the Station Image   15
   Promotion and Sales   18
   Sales Promotion and Client Relations   19
   The News Department and the Station Image   20
   Audience Maintenance   22
   The News Theme: A Community Link   23
   Music and Slogans as Theme   25

4 **Promotion Elements**   26
   On-Air Promotion   26
   The Promotion Mix   28
   Publicity and Special Events as Promotion   36
   Public Relations as Promotion   38
   Copyrights and Releases   39

5 **Planning the Promotion Campaign**   41
   Objectives and Goals   41
   Promotion and the Strategic Marketing Plan   43

6 **Promotions That Work**   51
   Getting Your Clients Involved   51
   Selling Promotion Ideas to Your Clients   53
   Creative Promotions   53
   Making Contest Promotions Work for You   64

7 Ethics and Promotion Standards   67
  Ethics in Radio Promotion   67
  When Is a Promotion Unethical?   68
  Parting Words about Contests   72

8 Radio Promotions and the Future   74
  Radio: The Growth Medium   74
  Trends   76
  Keys to Target Marketing for the 1990s   78
  Radio and the New Technology   81
  Promotion Directors and Departments of the Future   82

Glossary   85

Suggested Reading   91

Preface

There is a dearth of information on specific promotion and marketing techniques in radio broadcasting today. This book covers basic radio promotion concepts and practices common to all radio stations regardless of size and location. It exposes teachers, students, and promotion practitioners to new ideas and experiences in radio promotion as a management function. As a reference guide and supplement, *Practical Radio Promotions* gives the reader a chance to step back and see what has worked and is working in other places.

This book is designed to help students learn the practical applications of various promotion concepts. It should serve as a useful supplement to their course work in broadcast advertising and sales, public relations and publicity (broadcast), broadcast programming, broadcast management (radio), and other media management/operations courses. It should also help teachers of these courses. Radio stations will find the book helpful in developing and carrying out all kinds of promotions, by drawing on the tried and untried examples, as well as the specific experiences of existing radio promotions practitioners.

Practical Radio Promotions is committed to

- increasing the effectiveness of promotion professionals
- improving promotion methods, techniques, and strategies
- sharing the ideas of promotion professionals
- improving the quality of promotion instruction at colleges and universities
- increasing awareness and understanding of radio promotion

Practical Radio Promotions is for college-level (teacher–student) instruction and on-the-job quick reference use for professionals. Its style is direct without talking down to the reader or sacrificing stylistic quality. Chapters 5 and 6 are the how-to-do-it sections. They contain experiential examples from radio stations around the country that reinforce concepts, terms, techniques, and relationships discussed in the text.

Promotion is a growing arena for entry-level jobs for college graduates interested in programming, sales, production, management, and station/network marketing. In addition, many programming executives and promotion/marketing staff members who are currently employed in the broadcasting industry will find this book interesting. For them, this is a refresher course and a way to keep abreast of current economic and programming changes taking place in the industry.

In the following chapters you will find strategies for developing promotion campaigns and budgets, attaining cooperation of other members of the station management, and team playing and gauging the station promotion climate. Chapter 2, Promotion Management, describes various promotion activities, including public relations, sales promotion, and advertising.

You will learn from other promotion directors' experiences how to build station image and what is involved in on-air and off-air promotion. This book gives you a step-by-step description of how to organize the promotion plan of action, known as the *strategic marketing plan*. It establishes how this plan of action fits into the overall marketing plan for the station.

This book reports on promotions that have been tested and tried, in order to encourage readers (students and nonstudents) to see how some of these could possibly fit their particular circumstances. The book discusses station image and how to market your station with a look at client–station relations, community–station relations, and audience maintenance. It ends with a look at ethics in radio promotion and at the radio promotions/technology and the promotion director of the future.

Because this first book is devoted to radio promotion, I have tried to show the amazing things radio promotion can do, the illusions it can create, and the rewards it can offer to programming people, general managers, teachers, students, promotion practitioners, and the audience. When I speak of "rewards" I am referring to rewards that spring from innovative and creative use of promotions. Because my roots are in radio, I have devoted my research to that medium.

Introduction

Radio promotion usually is thought of as a method of gaining public notice for a station. That assumption is correct, but it is superficial. Public exposure is not the entire goal of a promotional effort. Any sizable market will have a number of radio stations competing for the available audience. Promotion is strongly geared toward *positioning* because of today's super-segmented radio market. Positioning is persuading the audience that one station or program network is really different from its competitors, a primary goal of promotion. This means creating and refining a station's image. It also implies adjusting that image for a specific market or demographic mix.

The way a listener perceives a station (the station's *image*) is sometimes more important than the station's programming (its *identity*). For example, a station that promotes "more music, less talk" may be perceived as playing less commercials, even though it may (counting its total daily output) carry more commercials than the competition. Successful promotion managers design campaigns to create these perceived advantages to ensure that listeners choose their station instead of the competitor's. The listener's perception therefore becomes central in promotion planning.

In *Radio in Search of Excellence*[1], Rick Sklar emphasizes the importance of perception when he says, "We radio people, promoting a medium that can't see, touch or feel, must keep a listener's perception to win listeners."

Promotion managers are able to learn about these perceptions through various research methods including *focus groups* and *call-out surveys*. Most information on the audience, however, is obtained from the controversial *diaries*. Many diaries are filled out long after a diary-keeper listens to the station. Trying to remember, when filling out the reports at the end of the week, the time and day that they listened makes reliability of this method suspect. Because of the delay between listening and reporting, many promotion managers believe the positioning of a station is more important for ratings success than actual station listening. Because diary-keepers are unlikely to report all of their listening accurately, Joseph Buchman in *Promotion and Marketing for Broadcast and Cable*[2], reports that many Arbitron diary-keepers and Birch respondents will respond in the survey by naming the station that presents the *image* with which they most closely identify, rather than a description of their actual listening behavior.

Radio today is in the business to provide extremely diverse audiences with various experiences. There are 10,000 radio stations in this country, reaching 99% of all households. Since 1950, contemporary radio's personal approach has resulted in a shift of the audience's application of the medium from family or group entertainment to individual companion.

Major radio stations and networks of the 1930s maintained standards of comportment that today would seem absurdly formal. Network announcers were expected to wear dinner jackets in the evening and to speak literate English with perfect diction. Both radio broadcasters and advertisers were sensitive to radio's status as a guest in the home.

In the early 1930s families gathered around their radios to listen to President Franklin D. Roosevelt's "fireside chats," a term used to suggest the informality, warmth, homeliness, and directness of those presidential radio reports to the people—a brand new phenomenon in American politics.

Physically, radio is more portable, than any other electronic medium. Today, people turn on the radio primarily when doing something else. The radio audience includes the commuter driving to work, the homemaker cleaning and cooking, the student doing homework, or the night security guard at work. Radio is in practically every home, room, and car. Three out of four adults are reached by car radio each week. In other words, radio isn't just at home anymore. Figure 1 is a graphic representation of how radio listening has changed.

Radio Reaches Upscale Consumers Where They Work
More than Half Have Radios Available in The Workplace

Household Income $50,000+	57%
Household Income $30,000+	59%
Professional/managerial men	53%
Full-time working women	64%
College graduate or more	57%
Major credit card users	63%
Homeowners	58%
Multiple car households	61%

Radio and American Workers: A Winning Combination
People of All Ages Have Radios Available at Work

Men 18+	59%				
Women 18+	63%				
Adults	18+	18–34	35–44	45–54	55–64
	61%	66%	58%	55%	53%

57% of workplace radios are employee-owned.
41% of workplace radios are company-owned.

On Weekdays 95.6% Of Americans Listen To Radio—Averaging 3 Hrs. and 19 Mins.
R=Percent Weekly Reach (Mon.-Fri., 24 Hrs.)
T=Average Daily Time Spent Listening

Persons 12+	R 95.6% T 3:19			Teens 12-17	R 98.8% T 2:26	
Ages		18+	18–34	18–49	25–54	35+
Men	R T	96.2% 3:30	98.7% 3:56	98.6% 3:45	98.3% 3:41	94.6% 3:12
Women	R T	94.3% 3:20	98.2% 3:29	97.3% 3:25	96.8% 3:27	92.0% 3:14

The Average American Listens To Radio For Almost Three Hours A Day
(Mon.-Sun., 24 Hrs.)

Average Daily Time Spent Listening

Persons 12+	2:59				
Teens 12-17	2:20				
Ages	18+	18–24	25–34	35–49	50+
Men	3:06	3:24	3:29	3:05	2:38
Women	3:01	3:00	3:10	2:58	2:55
Adults 18+ in households of		1–2 Persons		2:58	
		3–4 Persons		3:05	
		5+ Persons		3:11	

▶ *Figure 1* Statistics regarding radio listening habits at work. Courtesy Radio Advertising Bureau.

The increase in radio listening was both spontaneous and created: spontaneous in that experimenters and early program listeners were motivated to build or buy receivers, and created in that many more Americans had to be persuaded that the sizable investment of time and money was worthwhile—that radio was more than a fad. Until late 1920, just before KDKA's first broadcast in Pittsburgh, Pennsylvania (KDKA was the first commercially licensed station), all receivers were homemade. Some, like the crystal set, were cheap and simple. Virtually all sets needed earphones for listening, making radio a solitary pastime, although several sets of phones could be hooked into the same radio at one time. Even when loudspeakers were developed, the additional amplification they demanded was so expensive they were not widely used. In fall 1920, commercially manufactured radio became available, principally in large department stores. Radio was basically an urban medium in its early years. High cost and poor reception kept farm families from owning sets, although sales of radios increased rapidly when stations offered market and weather reports.

In recent years, technological innovation in receiver design is the single most important development that accounts for the ever-increasing popularity of the medium. Boom boxes, and walk-mans have boosted receiver sales over the $3 million mark annually. There are 12 million walk-mans in use. Radio's portability has made it very attractive to advertisers.

By means of research, it has become standard practice for a radio station to identify a significant segment of the available audience and to devise appealing programming for this target audience. Successful radio stations evolve gradually. Extensive planning goes into choosing what to offer the listener: Finding what will make your station stand out and give you the edge on your competitor. The principal content of a station's programming is known as *format*. The format is designed to appeal to a particular subgroup of the population, usually identified by age or socioeconomic characteristics.

In reality, few listeners probably know or care what name is used to describe the format of their favorite radio station. However, the selection of a name is important to management and the sales staff in projecting the station's image and in positioning the station for advertisers.

Practitioners know that both the size and composition of a given station's available audience may change throughout the day. There is no such thing as a "loyal" audience. Audiences stay tuned to your station as long as you continue to deliver programs that fulfill their diverse needs. Even when you do, they tend to hop from station to station. Therefore, your station must engage in creative image campaigns that attract and hold audiences. If your station is struggling among many to capture a specific demographic group you might find it expedient to change the station's focus and redirect efforts toward a more accessible, more easily available demographic group.

Radio formats change quite often as stations search for the "perfect" format. The station should detail strategies ensuring the success of such a campaign, as well as determine the resources required to do so.

While the emphasis and audiences for each format vary the basic elements of promotion are the same. Promotion is the activity that separates the successful from the unsuccessful station. These days, the economic gap is very large.

NOTES

1. Sklar, Rick. *Radio in Search of Excellence: Lessons from America's Best-run Radio Stations.* Washington D.C.: National Association of Broadcasters, 1985, p. 70.

2. Buchman, Joseph, G. in *Promotion and Marketing for Broadcast and Cable.* (Second Edition) Prospect Heights, Illinois, 1991, pp 139–164.

1
The Growth of Radio Promotions

THE GOLDEN AGE OF RADIO

Radio has captured the public's imagination since its inception. In its earliest expression, radio offered to free listeners from the insulation imposed by geography and an inadequate mastery of science. Any form of human expression that could be conveyed through sound was now available to the most isolated farmhouse and to the most urban city dwelling.

The Golden Age of Radio, from 1930 to about 1953, was full of new ideas. During this period, radio presented reformers and rogues, messiahs and maniacs, saints and sinners. In 1932, the United States inaugurated a president who promised a new deal to a citizenry burdened with economic depression. Franklin D. Roosevelt used radio to talk directly to the American people. During this period, a whole breed of outlaw stations developed in the Southwest, especially in Texas and Oklahoma. These stations operated without licenses because their owners said they transmitted intrastate only and so were not liable to Federal Radio Commission (FRC) jurisdiction.

But when most people speak of old-time radio, they mean the mainstream network entertainment programs developed in the 1930s. These radio formats stayed popular through the 1940s and into the 1950s. Most program types transferred successfully to television.

One genre that developed in the early 1930s was *comedy-variety*. A comedian acted as master of ceremonies to introduce and bridge the various acts and guests on the program. Often the comedian had come out of vaudeville. This genre initiated the radio careers of Eddie Cantor, Al Jolson, George Burns and Gracie Allen, Ed Wynn, Fred Allen, and Jack Benny—all performers who earned near-legendary status in radio. In the early 1930s, radio brought together a mixture of drama and news. Out of this came the "March of Time" (1931–1945), a radio program created by the weekly news magazine *Time*, in which several of the week's news stories were dramatized by actors. Radio stations broadcast contests and games, children's shows, public interest programs, and classical, light classical, Western, and popular music. There were sports broadcasts, religious programs, country music programs, disc jockey programs (DJ), and every kind of dramatic and music program imaginable.

It was magic and advertisers loved it. Listeners across the country could hear symphonies from the great music halls in Boston, or they could vicariously attend a big band performance at the Grand Ballroom of the Waldorf-Astoria Hotel in New York City. Radio exposed everyone to the largely white, Western European cultural and social mainstream considered representative of America.

THE RADIO NETWORKS ARRIVE

In 1922, David Sarnoff of RCA wrote a memo to his staff arguing that the novelty of radio was wearing off; to convince people to keep buying radio sets, better programs would have to be offered. In 1926, RCA formed the National Broadcasting Company (NBC) "to provide the best programs available for broadcasting in the United States."[1] Sarnoff's idea was considered revolutionary. It was so successful that NBC formed a second network a year later to accommodate increasing demand. The stations became NBC-Blue and NBC-Red, and coast-to-coast broadcasting became a reality.

Along with the need for programmers to meet popular tastes, there was increasing realization that radio's appetite for program material was enormous. Audiences waited anxiously to devour more programs. New entertainment forms had to be created to meet increasing audience demands. Radio needed programming that could attract listeners and keep them loyal to a certain sponsor or a particular spot on their radio dial.

As the Congress-Cigar-Company's 26-year-old heir, the late Bill Paley was fascinated by the role that radio advertising played in boosting his father's business. In 1927, Paley bought a 16-station network that had attempted to challenge NBC's dominance. His United Independent Broadcasters (UIB) became the Columbia Broadcasting System (CBS).

The profits earned by AM radio stations during World War II had not gone unnoticed. After the war, hundreds of people applied to build radio stations. Total time sales climbed from $176.5 million in 1945 to $275.6 million in 1948. However, the number of radio stations on the air was so large that average annual time sales per station actually dropped from $180,000 to $133,000 (Figure 2).

Popular Shows and Formats

The new networks were vigorous in their programming efforts, but NBC's two networks aired the most popular radio programs. NBC's greatest hit in 1929 was "Amos 'n' Andy." As radio's first situation comedy of sorts, the show centered around the adventures of a group of black workers, one of whom owned the Fresh Air Taxicab Company. The voices of Amos and Andy were those of two white men. Even within the context of that era, many black groups considered the show to be loaded with negative stereotypes.

By any measure, the most popular radio entertainment genres during the golden age were the mystery and action-adventure series. Among the most successful were "Gangbusters," "Calling All Cars," "The Fat Man," "Sam Spade," "The FBI in Peace and War," and "The Green Hornet." The action-adventure format has continued to be popular on television with shows like "L.A. Law" and "Magnum, P. I."

As with daytime television today, daytime radio in that era had its soap operas, including "The Romance of Helen Trant," and "Pepper Young's Family." For kids

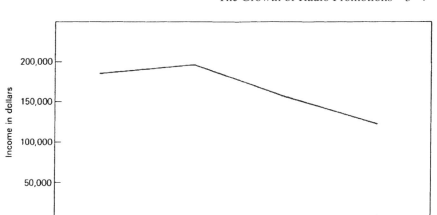

▶ **Figure 2** *Average radio station compensation 1945–1948. Courtesy of the Federal Communications Commission.*

there were "Jack Armstrong," "The All American Boy," "Superman," "Uncle Don," and, of course, "The Lone Ranger."

If all of these formats sound familiar it is because television borrowed so heavily from radio. Radio, on the other hand, borrowed from vaudeville and books.

Radio's significance as an arbiter of the national mood was brought home to many in 1938 when Orson Welles and *Mercury Theater on the Air* produced and aired the scariest radio drama, "War of the Worlds." Thousands who did not hear the opening announcer say it was just a drama panicked, believing Martians had invaded Earth.

In 1943, NBC's preeminence in network radio suffered a devastating blow when the FCC banned the operation of two networks by one owner. NBC sold its Blue network to Edward Noble who renamed it the American Broadcasting Company (ABC).

The Crossover Begins

In 1948 NBC radio stars defected to CBS (formerly UIB), encouraged by a bigger paycheck. CBS offered big-money contracts to Jack Benny, George Burns and Gracie Allen, Ozzie and Harriet Nelson, Red Skelton, Bing Crosby, and others. This helped establish CBS as the top network in the late 1940s and gave it a nucleus of talent for its new television network.

TELEVISION ARRIVES

In 1947, television began to grow at a phenomenal rate. Sales of television sets soared. The FCC was flooded with applications for television broadcast licenses. A freeze limited the number of stations to 108 (which was the authorized number before the freeze). The freeze prevented the full development and marketing of television. When the ban was lifted in 1952, radio broadcasters were shocked to see the rapid move toward television, and that television was invading radio's advertising market and listenership. Several media observers predicted that television's effect would be too devastating for the older medium to overcome. Many radio station owners sold their facilities to survive.

In the earliest days of television, there was no large backlog of syndicated programs to fill a station's schedule. For much local programming, stations used old feature films. These were the only films Hollywood had made available to its young rival medium, so television acquired an old-movie reputation. Some stations even ran silent films, adding narration and sound tracks. Some stations used short musical films—well-known bands, vocalists, and singing groups—and put them together in the form of a visual disc jockey show, anticipating a video format that Music Television (MTV) would reinvent 30 years later. At this time there was no videotape, so it was all filmed live.

Radio's Influence on Television

At the start, a network sometimes took a radio series and made an adaptation of it for television. Some series were simulcast, that is, aired concurrently on both radio and television. Soon, the first prime-time hit programs emerged. In 1948, the "Texaco Star Theater," a comedy-variety show, went on the air, launching the television career of Milton Berle. "Uncle Miltie" was so popular that he became known as "Mr. Television."

Television's Effect on Radio

By 1955, radio revenues had fallen to $90 million annually, and it was apparent that radio had to come up with a strategy if it was to survive.

The problem was television. Television began its meteoric rise in popularity in 1948. People bought television sets and deserted radio in droves. They peered at tiny screens and saw poor imitations of some of the programs that radio had already done so well. But that did not matter, because they also saw some of their favorite radio stars now on the screen. It was a bad time for radio. Average station billings were falling, network audiences were falling, network billings were falling, and the nation was entering an economic recession.

When television arrived, in 1948, comedian Gracie Allen supposedly observed that "nobody watches radio anymore." Many new television stations were established by wealthy newspaper and radio station owners. Radio lost a tremendous amount of revenue to the newer broadcast medium. Just as many people think the good old days of magazines are gone forever, some contend that radio will never be what it was during its golden age, the years prior to 1950.

Radio in the 1950s

A major change in radio broadcasting during the 1950s was that stations began to rely more on recorded music. The network feeds were replaced by the DJ who spun discs and filled space between cuts with informational chatter and sponsored announcements. Earlier radio programs, such as "Your Hit Parade" and "Make Believe Ballroom," provided the formula that music-oriented stations of the 1950s could emulate. Recordings of the popular big bands filled the airwaves along with the singing styles of Frank Sinatra, Patti Page, Bing Crosby, Nat "King" Cole, Sammy Davis, Jr., and others.

The prosperity of the 1950s allowed teenagers more disposable income than at any other period in U.S. history. They had grown up with mass communications—especially radio—and were receptive to phenomena that helped define their identity as a force to be reckoned with in society. Rock and roll (a term invented by DJ Alan Freed) was the perfect expression of the teenage culture. In 1955, Bill Haley's recording of "Rock around the Clock" struck gold and sold over a million copies, thus ushering in a new era in contemporary American music.

The phenomenon of this distinct and definable cultural group, combined with the affluence of the 1950s, made teenagers attractive as consumers. They had money to spend. Bill Haley and the Comets, Elvis Presley, Carl Perkins, Chuck Berry and the Everly Brothers had records to sell. Now the DJs were dealing with more than musical taste. They had to deal with a lifestyle.

Formats

Stations began emphasizing a particular sound in order to attract a desired audience segment (a target audience). The concept of *formats* had evolved. In fact, these stations decided not to compete with television for a general audience. Instead, their formats helped them to tailor programming to reach a specific audience. They then sold advertising time to companies that wished to reach that segment.

Prior to the arrival of rock and roll, soft drink marketers also had discovered the effectiveness of relating to the lifestyles of audiences in the "Pepsi Generation." They found this new rock and roll generation equally responsive to an appeal to having fun. Sales of grooming products, fast-food, and other goods were fueled by radio's successful recovery from its downward spiral through program formatting. Not all radio went to rock and roll. Some stations stayed with the adult audience. The companion phenomenon was not exclusive to teenagers: adults had discovered radio's utility as a companion. Where television could not reach or where full attention could not be given—in the car, at the beach, in the shower or backyard, on a fishing trip, in the streets or on line for the bus—radio still could entertain (See Figure 3).

Promotions

The *format* concept had caught on. But stations with competing formats were hungry for listeners. Stations had to make sure their format stood out; they had to create a particular image in the minds of their listeners.

Promotion was the answer. This became a conscious act of using persuasive information about the station, programs, or station personalities (on- or off-air) to generate increased listening and a positive image.

Stations engaged in either *specific* (tune-in) or *generic* (image) promotion. An on-air spot or print ad in a program guide that promotes a particular show on a particular day is specific promotion. Thus a promo that tells listeners to tune-in tonight at 8:00 to the "Brian Derrick Show" is a specific spot. In contrast, promotion and advertising that focus on the overall qualities of a station are generic or general image promotion. On-air promotions quickly developed: record giveaways; contests

Radio—The In-Car Companion

Cars with radios	95%
Used cars with radios	99%
Commuters' cars with radios	97%
Car commuters reached weekly by Radio	86%
Car travel as a share of all travel	82%
Adults with drivers' licenses	82%
Fleet and rental cars with radios	98%
AM/FM car radios	88%
Car radios with analog tuning	65%
Car radios with digital tuning	35%
Total car radios (millions)	131.4
Total truck/van/RV radios (millions)	37.8

Drivers Tune In To Radio

68% of adults listened to the radio every time they used their car during their last 10 car trips.

Percent of Driving Time Spent Listening Among Adults Who:	Adults 18+	Men 18+	Women 18+
Commute to work	97%	97%	97%
Shop for groceries	78	78	78
Shop other retail	81	81	80
Dine out: Fast Food	76	77	75
Dine out: Restaurant	76	79	74

3 Out Of 4 Adults Are Reached By Car Radio Each Week

Percent Reached In Cars (Mon.-Sun.)

	Adults 18+	Men 18+	Women 18+
24 Hrs.	78%	82%	73%
6 AM–10 AM	56	63	50
10 AM–3 PM	59	61	57
3 PM–7 PM	61	65	58
7 PM–Mid.	38	41	35
Mid.–6 AM	14	19	9

Radio: It Isn't Just At Home Anymore

Share Of Radio Audience By Location (Mon.-Sun., 24 Hrs.) Persons 12+

46.8% At Home
24.8% In Cars
28.4%* Other Places

	At Home	In Cars	Other Places
Teens 12–17	65.5%	18.8%	15.7%
Men 18+	35.8	30.6	33.6
Women 18+	54.0	20.4	25.6

*Other places includes listening at work and while shopping in stores.

▶ *Figure 3 Statistics regarding car radio listening habits. Courtesy Radio Advertising Bureau.*

and statements on contest rules; telephone calls taken over the air; fanfare for guest interviews and feature stories; timely tidbits of time, weather, and traffic (service information); games; ID and logo promos; theme promos; and on-air billboards of upcoming shows (a line-up of the evening's entertainment).

CONCLUSIONS

Although remnants of the golden age linger through radio formats that specialize in big band music, nostalgia, and the occasional television appearances of stars like George Burns and Bob Hope, the magic medium spins a different spell today. Radio has undergone tremendous changes. In 1990, there were approximately 10,200 commercial radio stations operating in the United States—4980 commercial AM stations, 3956 commercial FM stations, and 1265 noncommercial FM stations. An estimated 507 million radios were in use, or about two radio receivers in the United States per capita. Practically every automobile (95%) in the United States has a radio. According to Kenneth Costa, Marketing Vice President for the Radio Advertising Bureau (RAB) "there are twice as many car radios in use (approximately 123 million) as the total circulation (62 million) of all daily newspapers."[2] Further, seven out of ten adults are reached weekly by car radio.

The youth culture and rock and roll demonstrated convincingly radio's ability to live comfortably without the mass audience appeal that it had lost to television.

Radio Programming Is Diverse...
There's Something for Everyone

There are 4,975 commercial AM stations and 4,269 commercial FM stations licensed by the Federal Communications Commission operating in the United States (as of 12/31/89). 1,422 non-commercial stations can also be found, primarily on the FM dial.

A programming format usually defines the kind of audience attracted to a particular radio station.

Percent of AM, FM stations programming each format.

Format	AM	FM	Total AM/FM
Adult Contemporary	15.6%	28.5%	24.0%
Country	23.6	22.7	22.4
Religion/Gospel	11.9	3.8	8.1
CHR/Top 40	2.7	15.8	8.0
Oldies	8.4	4.5	6.5
Middle of the Road	7.7	2.2	5.2
Talk	6.8	0.2	3.9
AOR	1.0	6.2	3.1
Easy Listening	1.5	5.4	3.1
News	5.4	0.2	3.1
Spanish	3.2	1.0	2.2
Big Band/Nostalgia	3.4	0.7	2.2
Urban Contemporary	1.4	2.2	1.7
Diversified	2.3	0.8	1.6
Classic Rock	0.6	2.5	1.4
Black	1.9	0.7	1.4
Agricultural	1.6	0.8	1.2
Jazz	0.5	0.9	0.6
Classical	0.4	0.8	0.6

▶ *Figure 4* The most frequently used radio station formats. *Courtesy Radio Advertising Bureau.*

Advertisers were able to see the efficiency of targeting audiences they wished to reach and the effectiveness of directing a pitch to consumers who share common, exploitable traits.

This concept has brought prosperity to an industry that in 1950 had been written off by many as being on the verge of extinction. Today, we have an array of formats designed to attract various demographic groups.

Figure 4 is a listing of the most frequently used formats in radio today. There are a host of other formats, or subformats. Many are variations of those listed.

NOTES

1. Sarnoff, David. *The Image Empire*. Barnouw, Erik. New York: Oxford University Press, 1975, p. 24.

2. Costa, Kenneth. in *The Radio Station*. Michael Keith and Joseph Krause. Boston: Focal Press, 1989, p. 36.

2
Promotion Management

THE MANAGEMENT TEAM: THE PROMOTION DIRECTOR

Promotion in radio is a deliberate attempt to persuade people to continue to tune in to your stations. The overall responsibility for this important task is usually housed in a department headed by a promotion manager (PM) or program director (PD). At some stations promotion units are found in the programming department.

The size of the promotion department is determined by the size of the station, ownership structure, size of the market in which the station is located, the priority given by the station management to promotion, and the level of competition in that particular market. These factors also influence staff sizes, ranging from a one-person operation with secretarial assistance to a staff of ten or more that might include a writer, a graphic artist, a researcher, etc.

The PD is responsible for marketing the station and its programs to audiences and, in turn, for selling the audiences to advertisers. This responsibility involves developing and implementing a marketing plan that identifies target audiences and advertisers, as well as techniques and methods to be used in influencing those constituents.

The *marketing plan* is the product of discussion among the PD, general manager (GM), the program manager (PM) and sales manager (SM), and sometimes the news director (ND). The main responsibility of the PM is to carry out the plan. However, successful execution of the plan depends on the complete cooperation and willingness of the entire station staff.

The promotion department is involved in such activities as the following: scripting, editing, previewing, and reviewing films; making audio and videotapes; planning talent searches and making selections; scheduling radio, television, and print production; planning and running entire campaigns; evaluating and maintaining media relations and buying; supervising artwork and design; researching and budgeting for sales preparation and presentation; producing brochures; writing reports; conducting station tours; planning community tie-in programs; supervising gift-premiums and client-giveaways; assisting with community receptions; assuring accurate newspaper and *TV Guide* listings; developing feature stories; dealing with set problems (TV); and creating station logos.

Some stations do not give promotion responsibilities to a separate department.

Audience promotion may be carried out by the program department, for example, and sales promotion by the sales department. In small stations, the GM may play the prominent role in promotion. The following list emphasizes the PD and the promotion functions. Whether directly involved or acting chiefly in a supervisory capacity, the PD is responsible for a wide variety of promotion activities. They may include (1) audience promotion, (2) community relations, (3) station image, (4) sales promotion, (5) press relations, (6) research, (7) news and (8) general administration.

Audience Promotion

- scripting
- previewing tapes, music tracks
- scheduling
- directing the creative process
- maintaining quality control in spots and entire campaign
- selecting talents as needed (actors, actresses, announcers)
- maintaining inventory
- evaluating effectiveness of spots
- recommending and coordinating trade arrangements with radio stations.
- evaluating media (which media might be the most effective for your campaign?)
- determining size and frequency of ad (especially in newspaper advertising)
- researching costs and effectiveness of the campaign
- selecting locations (in outdoor and transit advertising)
- conducting post-campaign evaluation

The PD can devise other forms of audience promotion. The PD is only limited by the imagination. The PD must organize, arrange, compensate—with cash or through trade—and follow through to make these forms of promotion work. Here are a few examples: laundry stuffers; supermarket bags; tie-ins with department stores, retail outlets, book and record stores; painting the sides of buildings (with permission); skywriting; parades; fireworks; mailings; posters; and personal appearances.

Community Relations

Working closely with the community relations director (CRD) (if there is one), the PD:

- develops public service campaigns
- conducts station tours
- develops community tie-in programs
- helps to answer letters and calls from listeners

Station Image

The PD coordinates the way the station looks, and gives input for everything seen by the client and audience, including station stationery, order and billing forms, and how the parking lot and visitors' waiting area should look.

Sales Promotion

The PD must have a thorough understanding of research, sales, and marketing techniques to be able to write and supervise the layout and production of the following:

- customized sales presentations
- sales brochures
- general sales presentations using audiovisuals
- agency and client receptions (development of theme, selection of appropriate facility, budgets, catering, entertainment, invitations, and decorations)
- client giveaways (familiarity with suppliers, sources, and new and inventive gifts)

Press Relations

The PD must

- develop and maintain close press contacts (lunches and personal meetings)
- ensure accurate newspaper listings
- develop feature stories on unusual events, specials, new programs, visiting celebrities, station personnel.
- organize press receptions

Research

The PD should work with the research department to make sure that all research bulletins and published research materials are presented in the most professional and creative manner. He should be able to translate research statistics, especially ratings statistics, into clear qualitative concepts.

News

The PD works with the ND on all aspects of news advertising and promotion.

General Administration

The PD must

- prepare and monitor the promotion budget
- coordinate projects with other corporate divisions
- prepare monthly reports
- keep in touch with program suppliers for promotion materials
- cultivate reliable and reasonable suppliers for printing, photo, etc.
- keep updated bios and photo files on station personnel and programs
- keep track of salaries, raises, vacation, sick time for department employees
- help department staff with company procedures
- supervise the training of new personnel and interns assigned to the department
- keep track of competitor's promotional activities
- assist the general manager with intra-station communication
- stay abreast of legal restrictions on advertising

PROMOTION DIRECTOR: QUALIFICATIONS

The qualifications required of an effective PD are as numerous and varied as the responsibilities. The PD should have knowledge of marketing processes and functions, including

1. knowledge of audience sales promotion goals and techniques
2. understanding of the promotional tools available to the station such as
   - *publicity*: identifying the available sources for publicity and being able to develop new ones
   - *promotions*: knowing both on-air and off-air promotion methods and how and when to effectively use them
   - *advertising*: knowing the distinctive qualities of each advertising medium, as well as the array of media selecting and buying processes

The PD should be able to educe favorable feedback from advertisers and audiences; she must be skilled in interpreting and using promotion research data. The PD also is familiar with the professional services available from printers, professional promotions associations (such as Broadcast Promotion and Marketing Executives [BPME, see Glossary]), and advertising and public relations agencies. A successful PD adapts to changing circumstances within the station as well as to changing broadcast trends. He is alert to the programming and promotion inclinations of the competition. This individual is outgoing, a good public speaker, and ethical in promotion practices and in dealing with other station personnel, companies, and organizations outside the station.

THE PROMOTION BUDGET

Included in the planning of a promotion are cost projections. The PD's budget may be substantial or nonexistent. Stations in small markets often have minuscule budgets compared to their giant market competitors. Again, the intensity and extent of a promotion is dependent on market competition.

A typical promotion at an average-sized station may involve the use of T-shirts, public speaking, and hand-out materials (such as buttons, posters, bumper stickers, and caps), depending on the kind or depth of the promotion. Some stations may use billboards and television (additional costs that should be factored into any budget). Figure 5 is an example of a bumper sticker.

The cost involved in promoting a contest often constitutes the primary expense. Michael Keith and Joseph Krause[1] report that when WASH-FM in Washington, D.C., gave away $1 million, it spent $200,000 to purchase an annuity designed to pay the prize recipient $20,000 a year for 50 years. The station spent nearly an equal amount to promote the big giveaway.

How much then, should a station spend on promotions? In order to determine this, the PD must have estimates of the effects of communications allocation. He or she will have to depend on computer planning models to project sales response estimates.

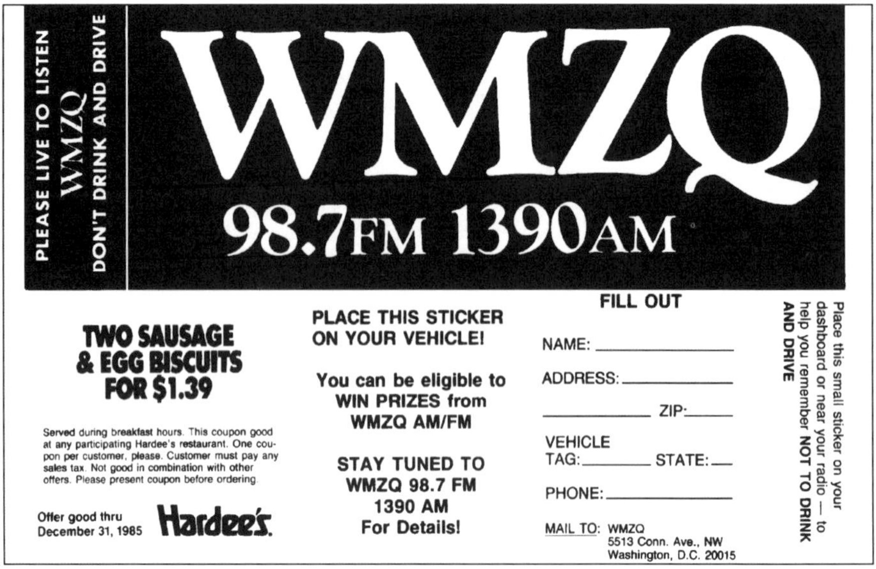

▼ 12   Practical Radio Promotions

WMZQ
98.7FM 1390AM

PLEASE LIVE TO LISTEN WMZQ
DON'T DRINK AND DRIVE

TWO SAUSAGE & EGG BISCUITS FOR $1.39

Served during breakfast hours. This coupon good at any participating Hardee's restaurant. One coupon per customer, please. Customer must pay any sales tax. Not good in combination with other offers. Please present coupon before ordering.

Offer good thru December 31, 1985   **Hardee's**

PLACE THIS STICKER ON YOUR VEHICLE!

You can be eligible to WIN PRIZES from WMZQ AM/FM

STAY TUNED TO WMZQ 98.7 FM 1390 AM For Details!

FILL OUT
NAME: _____
ADDRESS: _____
_____ ZIP: ____
VEHICLE TAG: _____ STATE: __
PHONE: _____
MAIL TO: WMZQ
5513 Conn. Ave., NW
Washington, D.C. 20015

Place this small sticker on your dashboard or near your radio — to help you remember NOT TO DRINK AND DRIVE

▶ *Figure 5*   An example of a bumper sticker that has coupons displayed on the peel off label. Courtesy WMZQ-AM.

Ideally, a station might choose to spend promotion funds up to the point when marginal costs equal marginal revenue. The biggest problem lies in the difficulty of calculating the effects of promotion costs on air-time sales. Also, the "payoff" from promotion takes quite a while to be realized.

The most desirable method of setting up a promotion budget seems to be the objective and task approach. This method is *goal*-oriented. The PD works with the station manager (STM) in establishing the promotion budget. From here, it is the PD's job to allocate funds for the various contests and promotions run throughout the station's fiscal period.

A statement of objectives forms the basis of a budget. The cost of relevant promotional tasks in a *promotion mix* are calculated. (The promotion mix consisting of four basic elements: advertising, audience promotion, sales promotion and station promotion). Assuming that the resulting figure is within the station's overall budget, it becomes the promotion budget for that particular campaign.

Cyndia Reynolds, Promotion Manager, WSZL-TV, Miami/Fort Lauderdale, is concerned about promotion costs and has come up with a concept she calls, "State of the Smart: Getting along without the New Technology.[2] The reality of producing high-quality promos that stand out in today's competitive marketplace is this: unless you work at a station with all the toys, bells, whistles, and state-of-the art gadgets, you have to do with what you have." According to Figure 6 you may not have a lot of money to spend.

It is bad enough competing with the other radio stations in your market that spend big bucks producing glitzy, award-winning promos. And with all the advertising messages from competing media bombarding listeners, viewers, and readers

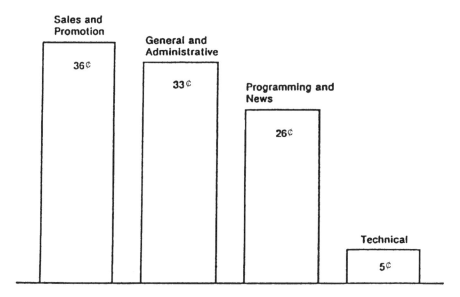

▶ *Figure 6* Allocation of typical radio station's expense dollar. Courtesy National Association of Broadcasters.

today, getting their attention isn't as easy as it use to be. The PD's job is made even tougher by this challenge. So, you have to think smarter when it comes to producing effective award-winning spots.

So how do you think smart and stay competitive with your competitors when they have the latest technological gadgets and you do not? Here are a few of Cyndia Reynold's suggestions.

1. Get attention—the audience should say, "This is for me."

2. Introduce the program or concept and focus the audiences on one central idea about what the campaign has done or will do.

3. Present convincing evidence—make certain that the message sounds as authentic as it is. Use language that people believe and understand.

4. Show how the product is in the listener's best interest.

5. Ask for action—be positive in tone, letting the audience hear the logic of the action and how it will bring the desired reward.

CONCLUSION

The role of the PD has evolved into a position of prominence. The job has become a profession and now demands special knowledge and advanced skills. PDs are often designated responsibilities such as creative services, advertising, or marketing to indicate their involvement in management decisions. This reflects the industry's emphasis on management theory and practices, goal setting, and research.

An important indicator of promotions influence is one's relative position in the station's personnel hierarchy. In the early years of radio and television, the PD frequently reported to (or was) the program manager. This person had little access to top management, especially at the larger stations. As station staffs grew larger and assignments more complex, a direct reporting line was established from the PD to the STM, and, in some cases, to the GM.

The position of PD is no longer a temporary assignment on the way up to a better position or a place to put the burned-out members of the staff.

Today, staffs of promotion departments range in size from only one to ten or more. In the small radio station (49 employees and under) promotion is usually handled by one or two people. This is the result of smaller budgets and fewer economic resources.

NOTES

1. Keith, Michael and Joseph Krause. *The Radio Station*, Second Edition. Boston: Focal Press, 1989, pp. 161–76.

2. Reynolds, Cyndia. "State of the Smart: Getting Along without the New Technology." *BPME Image*, October, 1988, pp. 12–13.

3
Selling the Station

DEVELOPING THE STATION IMAGE

Community Involvement

Image-building and station positioning are the primary objectives of audience promotions: running spots that tie station activities with the community. This will go a long way in helping establish station's value to the community.

Effective as programming strategy may be, it will not do any good unless the potential audience knows about it. Radio stations cannot depend on audience dial-switching and newspaper schedule listings to inform the public of new programs or time changes in old ones. Because a radio station sends out a continuous stream of images all day, a few words about the station's commitment to an image campaign are simply not enough to make any significant shifts in listener attitudes. If the words are not backed by action, they have no meaning whatsoever for the listener.

One definition of *station image* or station identity is the sum of *all* impressions of the station held by the audience. Every connection between you and your audience, as well as every interaction between a potential listener and someone who represents your station, influences the listeners' perception of the station. On this level and in varying degrees, everyone in the station has an influence on the station's image.

Primary factors influencing station image include

- *programming:* program selection/style (format), and network affiliation
- *news:* editorial policy (philosophy), new logos, anchor images, reporter images
- *promotion:* on-air promotion style, generic program promotion, general news promotion, television advertising, billboards, print ads, public relations, media relations
- *community involvement:* public service, public affairs, editorial positions
- *business image:* receptionist's voice, stationery, and business graphics
- *sales:* salespeople (attitude), sales promotion, commercial acceptance standards

Dealing with the image aspects of every one of these areas individually is not practical. Building a common focus for the station can produce the desired changes in audience perceptions.

This community-involving promotion has done very well for station WMAL in Washington, D.C. (Figure 7).

Community Involvement and Sensitive Promotions

Many radio stations shy away from controversial community issues, and rightly so. Wendy Drinnon, Associate Editor of *BPME Image Magazine*[1], gives examples of stations that have succeeded or failed by taking stands on political issues that don't necessarily concern the stations.

> As one station in Los Angeles found out, it just might make headlines and bring about positive publicity. However, in taking the risk, the station relied on a keen and intuitive knowledge of its listening audience, knowledge not necessarily provided by market research. Certain factors had to be considered as well.
>
> Los Angeles was still in a state of shock after an elementary schoolyard shooting several days before in which a young man with an AK-47 killed five children and wounded 29 others. Citizens no longer felt safe with such high-powered paramilitary weapons only in the hands of criminals and lunatics. Many chose to arm themselves more heavily. City Councilman Nate Holden saw the availability of the guns as the problem, and announced a $300 cash offer for every AK-47 or Uzi that was surrendered to Los Angeles police. The Councilman received criticism from the Mayor and fellow Council members who called his offer unrealistic at best.
>
> KIIS-FM/Los Angeles DJ Rick Dees thought the Holden idea deserved merit and support. In the same spirit, he proposed to air an offer of $1000 to the first person who would surrender an AK-47 or Uzi to police and present their receipt at the station. All other receipts obtained in a 24-hour period would receive $300.
>
> Within an hour, a young woman came to claim the $1000. She had talked her husband into letting go of his AK-47.
>
> Receipts continued to come in throughout the promotion and, by its end, KIIS received a total of 58 (one from a gun shop owner) plus numerous cash donations that were channeled to Mothers Against Gang Violence. Everyone sighed with relief. There had been no incidents and the solicitation had gone smoothly.
>
> Dees and KIIS had correctly read their audience.
>
> The following month, the Los Angeles City Council passed an ordinance outlawing the possession and sale of AK-47s and Uzis. Even gun control activists were surprised, for never before had such a stance against the National Rifle Association been successful.

This was a fairly successful promotion of a very sensitive issue. It may not always work that way. It is advisable to put your promotion to a test—especially sensitive promotions. Ask a colleague who has not heard anything about your planned promotion campaign to listen to your idea. Explain the promotion. How long does it take you to do so? Let your colleague ask questions about unclear

**WMAL AM 630 INVITES YOU TO
BE A PART OF WASHINGTON'S FUNNIEST TRADITION!**

GROSS NATIONAL PARADE

WITH WMAL'S TRUMBULL & CORE • SUNDAY • MAY 5 • 2 - 4 PM

JOIN THE CROWD!!
WMAL AM 630, Washington's favorite radio station, invites you to be part of Washington's favorite and funniest tradition, the **9th Annual WMAL's GROSS NATIONAL PARADE!** Follow WMAL's wild and witty afternoon personalities, **BILL TRUMBULL AND CHRIS CORE**, as they lead the most unusual procession of hilarious units down M Street between 18th Street and Wisconsin Avenue, N.W. in the heart of Georgetown.

SHED YOUR INHIBITIONS.
Discard your conservative veneer. Join marching bands, dancing chickens, corporate clowns, lawnmower drill teams and political look-a-likes. Here's your chance to poke fun at politicians, spoof current trends and entertain thousands of spectators who line the parade route.

IT'S HILARITY FOR CHARITY.
Proceeds from the parade are donated to the Metropolitan Police Boys and Girls Clubs to help prevent juvenile delinquency in the nation's capital. The Clubs maintain 11 different clubhouses throughout the District. These clubhouses provide counseling and programs for fun and learning that help youngsters stay out of trouble.

WIN CASH PRIZES.
First, second and third place favorites win $300, $200 and $100 respectively. Then there are also those famous **Wacky GNP Awards** such as the Peanut Award for the nuttiest entry.

PARTY WITH THE CROWD.
Cheer the winners, boo the judges at the GNP Awards BASH immediately following the parade.

TO ENTER
It's easy. Just put together a unit that will entertain the crowd. Be silly, be funny. Surprise Georgetown and the world with your wild and wacky parade entry. In other words, let it all hang out in the name of charity. Grab some friends, colleagues, roommates or relatives, fill out the attached entry form and send it in by April 15, 1991.

NEED HELP?
For more information or help with an idea, call: LaToya Woodbury or Amy Rosen at 202/686-3215.

▶ *Figure 7* An example of community outreach by a radio station. Sponsored by WMAL, the Gross National Parade helps the Metropolitan Police Boys and Girls Clubs. The GNP attracts hundreds of groups and organizations each year. Courtesy of WMAL-AM.

points. If you find it difficult to explain the promotion or that you have to give too many details to answer the questions, try to make things simpler and more practical.

Jack Macdonald, in *The Handbook of Radio Publicity and Promotion*[2], recommends associating promotions with community activities and events. If your promotions are logically associated with local events, they will have the extra support needed to make them real winners.

Before your station gets involved in a sensitive political community issue, be sure to consider these points. Whether the issue is gun control or saving a community landmark from demolition, it is imperative to have a solid understanding of your audience's viewpoint. Wendy Drinnon warns never to assume your listeners agree with your position. If there is the slightest doubt that you will not receive their full support, rethink your position. Radio station involvement in political community issues can be a hazardous way to enter the spotlight. However, like KIIS-FM's stand on gun control, the payoff often is worth the risk.

PROMOTION AND SALES

Have you ever noticed that people in the communications industry are often some of the worst communicators? Have you noticed how the concepts of territoriality, domain and compartmentalization dramatically narrow operation areas? Let's examine two important departments at a radio station—sales and promotion. These departments have identical objectives: marketing the station. Although they both work toward this common goal, they usually do not interact or effectively communicate.

Promotion tools are an integral part of the sales effort. Because of the sometimes frantic nature of the broadcasting business, people often narrowly focus on their own area. A narrow focus can often result in the failure of a station promotion campaign. One solution to the problem is direct communication. Part of each department's job is to tell the other what they are doing. Sales and promotion must meet periodically, share project information, and exchange potentially adaptable advertising and promotional materials. After all, advertising and promotion sell programs to listeners, while sales sells listeners to advertisers.

The sales department depends on promotion to deliver listeners and create the right station image for continued sales. The station depends upon sales to make a profit so the station can afford good on-air talent and produce and promote new programs.

Promotion budgets at radio stations are comparatively small (see Chapter 2, Figure 6). Effective promotion can lead to higher sales revenue, and higher sales revenue can lead to higher promotions budgets.

Obviously, the long-term benefits of a locked-in promotion schedule outweigh the short-term sacrifices that every unit or department of a radio station must make to ensure its success. After all, on-air promotion helps generate higher ratings, which in turn makes it easier for the sales department to sell airtime at higher rates.

Conversely, PDs should be more responsive to the marketing needs of the sales department. Sometimes the audience-oriented promotion staff often fails to understand how their materials will be used by sales, or how effective their efforts and materials can be in helping to market their station to the outside advertising community. Important tools such as trade ads, sales presentations, and one-sheet promotional hand-outs are usually cranked out quickly and sometimes have very little relation to the overall marketing position of the station. These tools are critically important in bringing new advertising clients to the station. A sales department must clearly communicate what tools they need to compel a client to listen. Such communication should clearly identify the target audience and the deadlines.

The promotion department should consult the sales department about its needs at budget time and prior to each sweep—the four week periods in each year when rating services gather audience data for all markets in the country. These sweeps determine network and station advertising rates, and they should include sales in plans for the station's overall marketing image. Sometimes additional materials are available from program distributors, which can be used as elements of a local sales kit.

Vicki Hoffman, Marketing and Advertising Manager of Broadcast Promotion and Marketing Executives (BPME), supports the need for cooperation between the sales and promotion departments. She observes that special promotions and community events are particularly important areas where sales and promotion can work together more closely. When an event or special promotion is being planned, she writes in BPME *Image* magazine October, 1988, p. 10,[3] sales involvement can help secure outside sponsors and prizes. Sales also can invite a client to the promotion department to create a special event tailored for that particular client.

Sales and promotion are thus interdependent; this is why they are both considered marketing departments. Revenue cannot be generated without promotion. Likewise, if radio stations were not sales-driven with a common bottom-line revenue, there would be no need for promotion.

SALES PROMOTION AND CLIENT RELATIONS

The goal of sales promotion is to give prospective advertisers an incentive to buy time on the station. Sales promotions that have an on-air component, such as a contest, actively involve the PD.

The client is a very important constituency in radio promotion. *Client incentives* that involve the promotion department are usually similar to strategies used in audience acquisition. Incentives include station *remotes,* which routinely draw large crowds to station-sponsored events (such as free concerts) or remotes that originate from a client's place of business. A station, in an attempt to create a good rapport with its clients, may offer free advertising (or discount rates) to clients who have the station tuned in at their store when account executives visit the store.

Another critical area in the image-development process is the use of research. Two kinds of data are needed to clearly understand current perceptions and locate possibilities for a unique station image:

1. *objective data:* cold, hard facts about what has happened or is happening, including ratings, news research, objective market surveys, etc.
2. *personal data:* the feelings, thoughts and viewpoints of people both inside and outside the station. Personal data is important because it reflects whether people will be interested in the image you want to project. To make a difference in audience perceptions, this perspective is critical.

Most promotion people think of themselves as creators and producers. The biggest problem they face is how to get everyone else to go in the direction they believe is best for the station.

THE NEWS DEPARTMENT AND THE STATION IMAGE

News and News/Talk stations can make the most of the uniqueness of their format and also create tremendous opportunities to stay in the public ear with just a little foresight, some creativity, a knowledge of local needs, perhaps some daring, and a willingness to dig in, take leadership, and do the work.

Radio news is promotable, whether you are a music station with a 2-minute newscast every hour, or a 24-hour all news format. Radio stations do not broadcast news because of FCC deregulation; they do so because it is essential to remain competitive. Listeners expect and demand it. News, therefore, deserves the same promotional support that radio stations give their music and on-air personalities. It is important to know the capabilities of your news department, your market, the competition, and your audience.

News promotion messages fit into two broad categories: *long-term campaigns* (using generic spots) and *short-term topical promotion* (using specific spots).

Long-term campaigns can be used by stations with low ratings to mount competitive assaults on the ratings leader in the marketplace, and ratings leaders can demonstrate bragging rights through long-term campaigns. Examples of short-term topical promotion include local story exclusives, special features on current issues such as crack-addicted babies and timely reactions to public events such as a court decision relating to women's rights. Short-time promotions require only one or two days of lead time.

The three common objectives of news promotion are to gain ratings supremacy, establish professional efficiency, and earn news-gathering credibility. These goals can be achieved by the hiring and firing of anchors, ensuring that you get the *best* talent available, changing newscast scheduling to make it more accessible and convenient to your listeners, and being factual and dependable with the news.

If traffic tie-ups are a big issue in your city and you're the only station with a helicopter, what you promote is obvious. Or, if you know your listeners want a clear, concise newscast with no fooling around, then design promotions to stress the professional no-nonsense approach of your news department.

But first let's focus our attention on a key player: your *news director* (ND). Without her cooperation and enthusiasm, any attempt to design promotions around the news is doomed to failure. As Joyce Kreig[4], Promotion Manager, KFBK/KAER/Sacramento suggests,

> Don't even think of asking your news director to do a commercial endorsement. Asking him to inject something into a newscast that he feels isn't a bona fide news event is also a no-no.
>
> You're on safer ground when you attempt to get your newspeople involved in your station's charitable or community service events. But

don't forget that strong sense of professional integrity at the hand of your news director. Asking him or her to sit in the March of Dimes dunk tank will send shivers of revulsion up and down his/her mic flag. Asking him or her to give a speech on media relations at the next March of Dimes board meeting will net more enthusiasm.

News promotion in the 1970s meant massive amounts of technology. From satellites to computers, a new age of electronic journalism had emerged—live news coverage, getting it fast, getting it first. Well, we've come a long way since then. However, it doesn't mean that we haven't looked to the past (those times when simplicity and reality were king) to build for the future. So, in the 1980s and now in the early 1990s, we work twice as hard to create promotions that are representative of the times. It is simple, it is warm, it is believable, and it is real.

According to Kreig, most of the news promotion we hear on radio stations seems to fall into one or more of the standard promotion categories[5].

Up-Close and Personal: an anchor in a non-news setting talks sincerely about his/her life, philosophy and/or family.
Comedy: usually done with weather, sports, or feature talent; puts them in a humorous situation, saying or doing something funny.
Team Promotion: the news team bakes cookies, rides horses, or, God forbid, puts together the news as a happy ensemble.
News in Progress: takes the listener behind the scenes to show what a hard-working bunch of journalists the news team is.
Emotional: makes the listener want to cry about something; comes in various flavors, bittersweet, touching, etc.; often involves children in some way.

There are other categories and subcategories, and some spots incorporate one or more of those previously listed, but you get the general idea.

Are there any *new* ways to promote local station news? There are basically two ways of thinking when answering this question. One school of thought says that there are no new ways, only new executions of old themes. Another says that there *are new ways*. We've all seen spots that were wonderfully imaginative visually or conceptually but left the viewer wondering what they were trying to say.

Part of the reason there seems to be a limited number of new ideas is that the goals for news promotion—the standard requirements, if you will—remain constant. They include the following:

- The concept has to fit the product. If your news product has always been solid and conservative, no amount of advertising will convince the viewer that your anchors are the hippest thing since "*Miami Vice.*"
- The concept has to be believable. Few things will kill your credibility faster than making claims that the viewer just won't buy.
- The concept has to offer some benefits to the consumer. Whether this is overtly stated, or subtly communicated, you have to make the viewer believe that there's something in it for him or her.

- The campaign has to enhance the product's image. The whole idea of good news promotion is to make the case for your product manager better than that of the competition.

AUDIENCE MAINTENANCE

Bumpers as Maintenance Tools

Radio stations with a high audience turnover rate (listeners changing to another station at the end of the quarter hour) may engage in various kinds of promotions to ensure that the audience stays tuned and flows through from one program to the next. Stations therefore use *bumpers* (on-air teasers that precede a series of commercial spots) to encourage listeners to stay tuned for upcoming programming. Bumper strategies include *billboarding*: highlighting the songs that will be played in the next quarter hour; promoting the next air-shift and its on-air personality; or intriguing the listener with information about an upcoming special, a news headline ("with details to follow") or a weather report ("in a few minutes"). These program bumpers can be used to keep the listener turned to your station from hour-to-hour and from daypart to daypart.

Contests

Highly effective contests are expensive, time-consuming, and require a lot of effort. On-air personalities must be very conversant with all parameters of the contest. Simplicity is crucial in any radio promotion contest. Don't limit your contest. Make it accessible to all. Make it practical and simple. Don't ask listeners with brown telephones to call in and win. This discriminates against those who do not have brown phones.

When KDKA, Pittsburgh, broadcast the presidential election returns on November 2, 1920, listeners to the historic broadcast were invited to call in and declare their presence. No prizes were offered; just being heard on the radio was prize enough in those days. The only winner that night (and a loser soon thereafter) turned out to be Warren Harding, who captured the presidency. Whatever small audience was tuned in that night later had the satisfaction of knowing that they had listened to what came to be accepted as the start of modern broadcasting. (The number of people who later *claimed* to be among that initial corps of listeners was much larger.) The pioneering election coverage triggered a listener response, and ever since then radio and television have capitalized on the public's willingness to respond, react, participate, and compete for prizes (everything from megabucks to dinner with a star, or passes for a movie).

From the earliest days, contests have been as much a staple of broadcasting as news, music, and sports. Well-conceived, well-run, well-promoted contests can achieve a great deal by involving listeners or viewers. They can get people talking about the station, the program, the personality, or the product being promoted. At their very best, contests can build an audience.

Bumper Stickers

As a community campaign, bumper stickers are effective if you can get drivers to use them. Some stations have vehicles that cruise the streets searching for vehicles bearing the station's bumper sticker, and they'll award cash on the spot. Many factors come into play in a bumper sticker campaign. A successful program must include the following: (1) a high-quality campaign; (2) a well-analyzed distribution scheme to determine the most effective way to reach the largest audience; and (3) an interesting contest or audience incentive that is linked with the distribution plan. Also, the bumper sticker should be of top quality. You don't want to suffer the wrath of listeners who can't peel the bumper stickers off their cars (heaven forbid they should ever want to). Also, the ink should be light-fast so that it won't fade after several months on the road. A bumper sticker should last a year or two.

Few people will put a bumper sticker on their car without an incentive. The stronger the incentive, the more participation. The more participation, the more success. If you choose to run a contest in conjunction with your bumper sticker campaign, think about tying it in with a local vendor. Giveaways build traffic (customers) for them.

Billboards

Billboards (outdoor promotion) are very helpful in reaching drivers, who make up a large part of the listening audience. Keep the billboard message simple. A motorist has only a few seconds to absorb your message. Transit promotions (on buses, subways, or trains) also are very helpful in reaching regular and potential audiences.

Speakers' Bureau

Another popular promotional activity to reach a station's community is the *speakers' bureau*, a public appearance by on-air personalities. Having your meteorologist speak at the local 4-H club or neighborhood luncheon can reach 100 people immediately. By extension, this could mean five times that number of new listeners for your station.

Remotes

Broadcast remotes are effective and they can be used for direct sales or strictly for promotion. Remotes are great for positioning the station's image in the minds of listeners. Broadcasting live from shopping malls, block parties, high school proms, or country fairs creates an indelible picture (image) in the listener's mind. Listeners get a chance to see, touch, and talk with this voice on the radio. This personal touch creates a sense of loyalty to the personality and to the station. The on-air personality becomes alive in the mind of the listener, and the next time they hear that DJ, they'll say "I know him."

Concerts

As part of its community outreach programs, radio stations sometimes stage concerts and other events. This requires careful planning and a willingness to take risks. A lot of money can be lost on a failed concert. There are all kinds of permits, clearances, and security arrangements to be made and/or obtained.

THE NEWS THEME: A COMMUNITY LINK

Almost every television station and some radio stations have a news theme or community image that they proudly promote in their local programming, community activities, and publicity efforts. A few of these include "We care for you," "Come home to . . . ," "The spirit of . . . ," "Turn to a friend," ". . . on your side," etc. News/Talk radio stations have special opportunities for creating responsible community promotions. In radio, the market is defined more by promotions than by themes. In television, the station is often identified by its image campaign. To be effective, it requires some action. Jack Hansen[6], founder and president of SPR News Source, Inc., a Minneapolis-based syndication company, is concerned that "irresponsible promotion is wide, varying from the simple attempt to buy listeners or viewers to downright criminal activity. And there are many examples of stations which simply don't deliver what their community image suggests. Their actions either don't support, or, in some cases, contradict their image.

The test is the same: Ask yourself, "Why am I doing this and why should my audience believe it?" If you don't like the answer, stop. If it's short-range, stop. If you doubt your station's ability to deliver, stop.

Local, national, and international news on News and News/Talk stations often look to local public interest as a way of calling added attention to their unique formats. Marketing the format and the station involves designing special community events and activities.

Jan Cromartie, Marketing/Promotion Manager, KFWB News 98, Los Angeles, says that KFWB News 98 is commited to serving the community by taking extra steps beyond just reporting the news. "We do this without violating our commitment to providing listeners with 'All News, All the Time'" (p. 14–15).[7] The station inserts public service-oriented program segments into the time slots not ordinarily occupied by news, concentrating on nonbroadcast activities.

Cromartie emphasizes call-ins programs as a way of getting people involved. Through special programs and program segments, listeners can get directly involved in community events and civic issues. Call-in programs can feature important civic leaders, notable guests, or talk show hosts addressing current issues. News and News/Talk stations can use informational programs to get involved with the community and strengthen their positions as community leaders.

KFWB, Los Angeles, designs call-in segments to cover topics such as tax tips, legal matters, and earthquake preparedness—a subject about which the station has become a recognized local authority. Such community-involving programming provides important public services, and it's programming the KFWB listeners have come to expect. In your town it might be hurricanes, tornados, drought, or ongoing unemployment.

A station can also create and promote a scholarship program for economically disadvantaged high school seniors. Such a program would be an ongoing message that the station cares deeply about the community's future. Involve your editorial board, the mayor of your city, and representatives of several top-level universities in the area. This, naturally, will generate more widespread support and public interest. Education is an important issue with a long shelf-life, and a scholarship program will say to your listeners, "We care about the future of our young people, your children." Involving the public in bettering the communities is really what community service campaigns are all about.

MUSIC AND SLOGANS AS THEME

According to Gary Graff, a music writer for the Detroit Free Press, the easiest way to test theme lines or slogans is to put them on large poster cards and show them to a panel from the front of the room. Determine the potential positioning options (work with station management, your advertising agency, and your market research company), then examine the various themes and slogans you might use to tie in your advertising and establish your position.

To use a simplified example, you might work at a station whose main market differentiation is playing more rock music with fewer interruptions than the competition. Once all parties agree that this is the position you want to establish through advertising, you can work with the advertising agency (or your own staff) to come up with three theme lines that effectively communicate the point. Put the three slogans on poster cards in big letters and bring them to the panel for testing.

If you are not sure which of two positions is best for your station (i.e., should we be more music or better music?), you should review more comprehensive market research. Monitor audience reaction by testing the three more music themes versus the three better music themes with an in-station panel.

When a station uses a theme to differentiate itself in the market, William Bartolini suggests,[9] it must remember that like any other advertiser it must *deliver* what it promises because viewers will have expectations based on that promotion. This is much easier to do if there is substance to the image promotion, and if the promotion is followed by action. If your actions speak louder than words, the whole community is more easily convinced, by observation, that you are delivering what your image has promised.

Image and theme selection is very important. Make sure the theme clearly states what you mean. Use one identifiable theme at a time. Whatever you do, do not follow the example of one midwestern top-15 market station. They had three separate theme promotions running at one time. Guess who was confused?[9]

NOTES

1. Drinnon, Wendy. "Strong Radio Sessions Hightlight 1989 Seminar" *BPME Image*, June/July, 1989, p. 20.
2. Macdonald, Jack. *The Handbook of Radio Publicity and Promotion*. Blue Ridge Summit, TAB Books, p. 16.
   3. Hoffman, Vicki. *BPME Image*, October, 1988, p. 10.
   4. Krieg, Joyce. *BPME Image*, November, 1988, p. 30.
   5. Krieg, Joyce. *BPME Image*, November, 1988, p. 33.
   6. Hansen, Jack. *BPME Image*, June/July, 1989, p. 34.
   7. Cromartie, Jan. *BPME Image*, August/September, 1989, p. 14–15.
   8. Graff, Gary. "Public Service Promotion" *BPME Image*, June, 1989, p. 42.
   9. Bartolini, William. "Quick Marketing Tune-up" *BPME Image*, June/July, 1987, p. 84.

4
Promotion Elements

A promotion must be compatible with a station's programming. You do not give away water beds on a Beautiful Music Station or symphony tickets on a Top-40 Station. These prizes would miss their intended demographics, since 60-year-olds are not big purchasers of water beds, and teenagers are not noted for their love for classical music.

ON-AIR PROMOTION

There are two primary categories of station promotion: on-air and off-air. On-air promotion is the most common promotion. The PD's challenge is to effectively market a station to expand and retain listenership. A number of promotional devices are used. Among them are call letters. Call letters convey the personality of a station (discussed in more detail later in this chapter). Call letters should telegraph instant recognition. They should have a retention element. When they don't, most stations tie them in with their broadcast frequency, such as WHUR-FM 96.3 or WXKS-FM 108. (The call letters are grouped with the frequencies.) It's never the call letters or frequencies by themselves. They become *one phrase*.

Slogans also improve the listeners retention factor. They are used as on-air ID, for example, "Music Country WSOC-FM, Charlotte, or "WTOP-AM, All News Radio." Slogans are positioning tools that enhance the station's image. They help fix the station's name and image prominently in the minds of listeners.

Since a radio station decides how it will use its airtime, promotional announcements or promos are supposed to keep the station's call letters and frequency in the listener's mind.

Promos remind regular listeners of the benefits of listening, and should capture the interest of those listening for the first time. Station slogans are usually tied in with promotional announcements. The following are a few of the ways to use promotional announcements on the air.

Personalities

Disc jockeys (DJs), news, and sports anchors may be promoted on behalf of the station.

Format

This amplifies the ways in which the station satisfies its community needs. This too could be tied in with the station's slogan. There are dozens of formats but all can be placed in one of the following categories: (1) music (2) information and (3) specialty. The music format is the most common among commercial radio stations. It is sometimes difficult to describe particular station's format in a few words because formats are fragmented, and some artists fit in more than one format.

1. The major music formats used by radio stations are
   - Urban Contemporary (UC)
   - Album-Oriented Rock (AOR)
   - Adult Contemporary (AC)
   - Beautiful Music (BM)
   - Classical
   - Nostalgia
   - Middle of the Road (MOR)
   - Country or Country and Western (C & W)
   - Jazz
   - Contemporary Hit Radio (CHR)

2. Information: There are two basic information formats and a third consists of a combination of the two. They are classified as
   - All-News
   - All-Talk
   - News/Talk or Talk/News

3. Speciality: There are many specialty formats. However, the following are the most common:
   - ethnic
   - religious
   - comedy
   - variety

Slogan
A slogan reflects the station's format and is usually tied in with the station's official or legal identification at the top of the hour as required by the FCC. The slogan also may showcase the station's role in the community.

Programs
The station also can promote itself through its programs. These promotions may deal with specific aspects of the program, personalities, upcoming programs and their special attractions.

Contests
Contests also are a big draw. Chapters 5, 6, and 7 give a few examples of contests that are simple and inexpensive to run. In deciding whether a contest has possibilities, ask the following questions:

- Is it cost-effective?
- How much time does it take? Is it all worthwhile?

- Will it distract listeners from regular programming?
- Is it exciting?
- Is it limiting? Can every listener participate?
- Is it timely? Don't promote baseball during the football season.
- Is it simple? Are the rules easily understood?

Warning! Make sure your contest is *legal*. Make sure it is a contest and not a lottery. Broadcasting a lottery is illegal and, therefore, a crime. There are three elements in a lottery, and all of these elements must be present for the promotion to be considered an offense: (1) a prize; (2) chance; and (3) some form of consideration. A prize is something that is offered and won by the contestant. Chance is present when the winners are chosen by a drawing and not because of the contestants' skills. Consideration is present if (1) the contestant must pay a fee to enter the contest, or (2) must possess a certain item for eligibility, and (3) if time is a factor for entering the contest. This determination differs from state to state. Off-air promotions are discussed in the next section, Promotion Mix.

Contests must be relevant to the lifestyles and interests of a station's audience. An effective contest, whether long or short, should involve the entire audience—this means both active and passive participants.

The most desirable promotion effort includes a total campaign with one unified theme. All promotional messages tie into this theme, rather than conflicting with it. One method is to divide the year into four quarters and let those quarters serve as your main operating modules. For each module develop an on-the-air theme. For example, during the winter months you might want to adopt a theme that ties in with the call letters of your station, such as, "WLBC, keeping you warm all winter." In the summer months you might try something like this, "WLBC, for summer fun." You might find it more beneficial to sub-theme each month. Your station identification spots should be tied in with your themes. Your promotion activities during these quarterly modules should reflect the ideas behind these themes. Remember your call letters are an important part of your promotion. Use them frequently and effectively. Take every opportunity (without sounding redundant or overbearing) to display your station's call letters and frequency. Some of these elements are on-air and/or off-air promotions.

THE PROMOTION MIX

The promotion mix consists of four basic elements: advertising, audience promotion, sales promotion, and station promotion. These elements are defined as follows.

Advertising (Usually Off-Air)

Advertising is any paid form of nonpersonal presentation and promotion of ideas, goods, or services by an identified sponsor. Advertising in this case refers to the buying of time and space in other media. Stations sometimes offer other media advertising time on their stations in exchange for time and space in the other media. In audience promotion, the major advantage of advertising is that the station has control over when and where the content will appear. This method can be expensive, especially if the station decides to advertise on television. Among the advertising media that radio stations use are newspapers, billboards, transit, magazines, broadcast, and news and features releases.

Newspapers Advertisements are often placed in a newspaper's TV and radio entertainment section. If a special program is being promoted, it may be placed in another section. Many radio stations use newspapers to advertise news specials and other informational programs.

Stations list broadcast schedules in daily newspapers and Sunday supplements. Radio stations do not use complete listings as widely as television stations. However, many radio stations do publicize their special programs, features, guests, or the subject of interviews or call-in programs.

Billboards Billboards are most useful if they are located in areas with heavy traffic—especially if slow-moving traffic. Many stations find billboards most effective because they can stimulate drivers to take *immediate* action by tuning in the station whose call letters and frequency are displayed. Billboards offer advantages as an advertising medium. They are available in various sizes, and they can be illuminated so that their impact is lasting. (See Figures 8 and 9.)

A radio station's outdoor advertising and promotion decisions will depend to a large extent on the importance of drive-time radio listening in that market. The much repeated cardinal rule for billboards is to keep the message *short*; three to five words make the most effective message. Make the message *provocative*. Colorful call letters and station frequency are not enough to create changes in listener behavior. Information from the rating and research organization can help you determine the most effective geographic areas for your outdoor advertising. It is advisable to target your outdoor placement to areas with good signal coverage but low listenership among the station's target demographics.

▶ *Figure 8* This billboard was designed to reinforce the news theme of WDAM, "News Seven, Eagle Eye on South Mississippi," and to show support for the university's baseball program. The baseball team is known as The Golden Eagles. Courtesy WDAM.

▶ *Figure 9* This is a transit campaign calling attention to the diversity, uniqueness, and special nature of WCCO Radio. Courtesy WCCO.

Although much of the outdoor media remains unmeasured, it continues to be a popular and inexpensive method of communicating with radio and television audiences. It was first embraced by radio stations because it was the medium (other than the station's own on-air promotions) closest to the point of purchase advertising—commuter travelers who would listened to the radio as they traveled to and from work.

Creatively, the sky is the limit for outdoor advertising and promotion. It can be fanciful, as in Volkswagen's classic billboards that featured the VW bug with the headline, "Relieves gas pains," or more dramatic, as in the now-famous Nike campaign featuring famous basketball super star Michael Jordan in which the outdoors was one of the primary media.

In the past, billboard advertisers had to rely on traffic studies, sometimes conducted by the same companies that were selling billboard space. A few years ago, the industry established an independent organization, the Traffic Audit Bureau, that measures and evaluates billboards for approach visibility, traffic flow, and a number of other factors. The Traffic Audit Bureau enables the Institute of Outdoor Advertising to calculate, for example, that a #50 outdoor showing in the course of its 30-day exposure reaches an average of more than four of every five adults (81.2%) in a typical market.

By comparison, these circulation figures would put billboards behind the average network-affiliated television station, which reaches 97% of the market every week, but ahead of newspapers which reach only 43% of the average market.

Here are some quick keys to success in outdoor advertising:

- simplicity is the cornerstone of effective outdoor advertising
- short copy is more attention-getting
- short, active words are more memorable
- simple typefaces are easier to read
- an uncluttered layout focuses viewer attention where you want it

When working in outdoor, there is help available to you for the asking. Your first source should be: The Institute of Outdoor Advertising, 342 Madison Avenue, New York, New York 10173. Most of the larger outdoor companies have creative directors, who will give free advice on creative approaches, or who will review your creative concept without charge.

The following examples demonstrate how, in outdoor advertising/promotion, you are limited only by your imagination. There are dozens of companies across America producing things that float and fly in the sky. For an idea of prices and concepts, you may want to contact Pie in the Sky, 1050 Charter Street, Redwood City, California 94043, (415) 366-7700. Pie in the Sky manufactures helium and col-air inflatables—including hot air balloon replicas.

Before embarking on your outdoor campaign, try to answer these questions:

Does the outdoor represent some truly imaginative thinking?
Will it take advantage of the bigness of thinking that must go into outdoor?
Will you use big, bold headlines (at least 20" tall), and big, eye-stopping illustrations or graphics?
Is it likely to stand out amid the competition?
Is the color bold—and does the background complement the message rather than competing with it?
Will the selling message register quickly from a distance?
Is the selling idea unique?
Is outdoor the right medium for the message?

Transit Radio stations place their advertisements on the outside of buses and trains, which travel great distances each day. Drivers, pedestrians, and riders see these advertisements. Advertisements placed in the *inside* of buses and trains are seen only by riders. They are effective because they have very little competition for attention. Platform posters are seen by a large number of commuters.

Taxicabs operate in most communities and advertisements on the rear or on the rooftop are seen by drivers, pedestrians, and people working in tall buildings. Other transit advertising media include display cases at airports, on the walls of bus shelters, and on bus benches.

Magazines Many radio stations operate in small markets where no general-interest magazine is published. However, large-market stations usually have access to city magazines and can use them to promote their format, programs, and image.

News and Feature Releases Most stations concentrate their publicity efforts on print media. These efforts emphasize the preparation of publicity materials. Among the more common materials are news and feature releases, which are stories about the station and its employees, including stories about new personnel, a new transmitter, awards won by the station or a staff member, and rating successes.

The best chance for having your material published is with local print media. If you believe your story has national merits you may want to try the national trade publications. When suburban papers, weeklies, and monthlies are added to the diminishing number of daily newspapers, no American market lacks a press corps.

The trade publications that may use local-market stories vary in the quantities they use and whether they use only radio, only television broadcasting, or only cable. The following publications might show an interest in your news release.

- *Advertising Age* (Crain Communications, Chicago)
- *BPME Image* (Broadcast, Promotion, and Marketing Executives Association, San Diego)
- *Broadcasting* (Washington, D.C.)
- *Communications Daily* (Warren Publishing, Washington, D.C.)
- *Radio Week* (National Association of Broadcasters, Washington, D.C.)
- *USA Today* (New York)
- *Variety* (New York)
- *Washington Post* (Washington, D.C.)

Unlike the trade press, newspapers are occasionally hostile to broadcast or cable press activities, perceiving them as rival media. Generally, however, reporters and editors are fair but human, requiring those considerations advocated in the five magical public relations concepts; friendship, optimism, credibility, knowledge, and momentum. Make sure to give editors stories they can use. Observe the following basic rules for effective press releases. They may sound quite obvious, but many publicists and promotions people forget them.

- Adopt a straightforward news style in keeping with the publication's writing style.
- Eliminate all the adjectives that belong in promotional brochures and advertisement.
- Double space the text.
- Use only one side of a page.
- Indicate when the information was released. Use a dateline.
- Include a contact name and telephone number of the person who created the release.

Photographs   When appropriate, photographs are included in news releases. A photo of a new transmitter tower or new remote truck may be considered effective publicity devices.

Press Kits   When stations wish to publicize special programs, they sometimes prepare a *press kit*. Press kits may include releases, photos of a new on-air personality, a new facility, an event in which the station took part, such as the station manager involved in some community outreach project. Other helpful materials are included in such a kit. The kit can be anything from a miniature suitcase to a folder. It usually holds about a dozen pages of typed information on letterhead and several glossy (black and white) photographs of station personnel and station facilities. Some press kits include large posters and even small customized merchandise (bumper stickers, a transistor radio with one dial position, pens, key chains, etc.)

News Conferences   Publicity events also attract the attention of the media, especially newspapers. News conferences are staged by many stations. The radio station invites reporters and photographers from all media to a conference to discuss or disclose significant news.

Miscellaneous Promotions Other advertising media used by radio stations to advertise their programs include ads at conventions, ballparks, malls and shopping centers, and on the sides of trucks, vans, and buildings, and skywriting, and direct mail.

Audience Promotion (On-Air or Off-Air)

Audience promotion is a deliberate attempt to increase the number of listeners tuning in to your station at any time. It also attempts to bring in new listeners. This kind of audience recruitment is advantageous during rating periods.

Promotions are on going yearlong events. Some promotions are a spontaneous result of some unusual happening, but you cannot plan a promotion based on chance. It is advisable to plan your promotions on a yearlong basis and let the spur-of-the-moment events serve as an unexpected gift—a gratuity. This enables you to set up your promotional activities and objectives based on a logical progression.

Throughout the year you will *normally* run a number of promotions. Yearlong promotions involve running a contest that culminates in the selection of a winner during the slow period. Throughout the year you should sponsor regular special holiday promotions (Christmas, Labor Day, George Washington's birthday, Independence Day, Valentine's Day, National YMCA Week, Groundhog Day, National Crime Prevention Week, Weight Watchers' Week, Future Farmers of America Week, St. Patrick's Day, April Fool's Day, Easter Promotions, Mothers' Day, Secretaries' Day, Fire Prevention Week, Be Kind to Animals Week, Baby Week, etc.)

Audience promotions have become very sophisticated. WKDD in Akron, Ohio, gave away Porsche automobiles and $10,000 credit sprees. Some promotions have been extremely successful. WHMD-FM in Hammond, Louisiana, doubled the size of its audience with a "Wheel of Meat" steak giveaway.

Image-building and station position are the primary objectives of audience promotions. Run spots that tie your activities with that of your community. This will go a long way in helping establish your station's worth to the community.

The methods to accomplish audience promotion goals are limited only by the imagination of the PD, the station staff, and the available budget. However, there are four major methods: (1) advertising (discussed previously); (2) publicity and public relations; (3) on-air and off-air promotions; and (4) public service. In most campaigns, a combination of methods is used.

Publicity, in this context, means the information in print publications or airtime on other broadcast stations for which the originating radio station makes no payment. Publicity, advertising, and other station activities that influence public perceptions of the station may be considered *public relations*. The emphasis is on personal contacts between the station and its numerous publics.

Sales Promotion (On-Air or Off-Air)

Sales promotion seeks to encourage the purchase of time on the station by advertisers and advertising agencies. It sometimes involves a cooperative effort among the retailer, suppliers, and the station.

The aim of advertisers is to communicate their messages, in the most economical way, to the demographic group(s) most likely to use their products. Sales promotion, therefore, attaches great importance to the broadcast media's ability to reach targeted demographics at competitive costs. The strength of a radio station is measured by its *audience reach*—the quantity or quality of its audience. Quantity in this case refers to the number of people who listen to that station, and quality denotes the characteristics of the audience, such as age, sex, and socioeconomic status.

Cost is an important part of the advertising equation. How does your station compare to competing stations as well as other media? The aim, using the station's biographic sketch, antenna power/reach, rate card, listener profile, and program format, is to convince the advertiser that the station can reach the number or kinds of people targeted by the advertiser in a cost-efficient manner.

An effective sales promotion campaign is the result of careful planning based on the following strategy elements:

- *Campaign purpose*: Is the primary purpose to project station image or to sell time?
- *Target clients:* Will the campaign be directed toward existing advertisers and time buyers, potential new clients, or both?
- *Client benefits:* What benefits will the campaign stress? What are the demographics and the cost of reaching them? Is there exposure to potential new customers? Other benefits?
- *Promotion methods:* What about the possibility of a joint promotion with an advertiser or advertisers? With the *vendor plan*, a particular radio station ties together a campaign for a retailer and then approaches several vendors who supply products to the retailer.

The Vendor Plan

One recent and highly successful effort involved a station in a major market that put together a vendor plan based on a Superbowl promotion for a convenience store. The package involved features on the upcoming game sponsored by the convenience store and included some promotional items distributed by the store. Vendors who supplied the store (soft drinks, potato chips, and other such items) financed a good portion of the campaign. The store in question achieved a 16% increase in sales over the monthlong campaign, and the retailer paid much less than if he had financed the whole plan without the aid of the vendor's contribution.

This kind of cooperation is cost-efficient and can reach targeted demographics—those who use the product or service the most. Other things to consider include the following:

- *Budget:* What costs will be incurred? Can they be justified by expected returns?
- *Scheduling:* During which quarter of the year will the campaign benefit the station most? Will that period be beneficial to the advertiser?
- *Program Impact:* Can the campaign be used to draw additional listeners to the station? Will it detract from programming?
- *Evaluation:* Will the campaign be evaluated on the basis of dollars generated or will other criteria be used as well?

Sales promotion targets the people who make decisions about the purchase of commercial airtime. This means advertisers, station representatives, and media buyers in advertising agencies. The station emphasizes what it can *do* for the advertiser.

The vendor plan is a cooperative sales promotion concept. Other ways to work promotions with stores include various merchandising methods such as aisle displays. Such a display may feature a combination advertisement for a soft drink and a radio station. In fact, such merchandising devices are typically an incentive for the advertiser to purchase airtime in the first place. Merchandising services are an added bonus to advertisers. The vendor plan involving the convenience store and a Superbowl promotion included a trip to the game for the store's owner. Inviting store owners to sports banquets or other appealing perks also help sell advertising.

Radio promotion on other media often can involve a trade—a swap of advertising time on the radio station for goods or services, sometimes including advertising time or space in other media. Trades frequently are used to supply prizes for contest promotions. Aside from traditional sales tools, printed collateral produced by radio stations take many creative forms. WHEB-FM, New Hampshire, publishes its own summer guide featuring station advertisers.

Celebrity Appearances Radio stations try to obtain maximum publicity when celebrities visit their community. A news conference, appearances in public places, and receptions are all publicity vehicles. The celebrity may agree to be interviewed on the station and might even tape a station promotional announcement.

The radio station has no control over the use of promotional materials sent to the media.

Broadcast Broadcast advertising is considered on-air station promotion because it refers to advertising done on television or on other radio stations.

Using audience demographic information for specific dayparts, a radio station can direct its television advertising to the kinds of people most likely to listen to its format or the program being promoted. The television medium's combination of sight and sound, motion and color can be very persuasive and could motivate to sample the product.

Station Promotion
Call Letters as Promotion Call-letters are a station's trade name or the logo most successful stations strive to enhance and protect. For radio and television stations, call letters, as much as anything over the years, have produced the brand awareness and consumer recall so critical to success in the rating books—which GMs translate into station revenues.

Call letters help to position your station in the minds of your listeners. Positioning is a concept that recognizes the problem of getting heard in an over-communicated society. It starts with a product, a service, an idea, your station, or even a person. The idea, thought, or concept is saturated to the point that it takes precedence over all other concepts, thoughts, or ideas related to the subject. It becomes second nature. For example, Xerox® and Kleenex® have positioned their products so well that they have become household words. Making copies is now "making a xerox . . . ," and we say "please hand me a Kleenex," instead of tissue. In the same way, stations try to position themselves in the minds of their listeners. So that when they think of listening to the radio, Station XXX is all they think about. To them XXX becomes the *only* radio station on their minds—and on their dials.

Positioning has changed the way radio market is being approached today. The basic approach is not to create something new and different, but to manipulate what's already in the listener's mind—to retie the communications that already exist. Imaging and positioning play a significant role when station sales personnel call on potential advertisers. The stiff competition among advertising media makes a powerful presentation essential. A client's familiarity with the call letters of you station makes selling your station's airtime a little easier.

Promotion-conscious broadcasters of the pretelevision era were just as dedicated to winning a large share of the audience as we are today. Stations used whatever was at hand to capture the listeners' attention. Call letters were configured in such a way as to convey a particular sentiment or meaning: WEAF/New York was "Water, Earth, Air, Fire;" WOW/Omaha was "Woodmen of the World;" WIOD/Miami meant "Wonderful Isle of Dreams." Placards were affixed to vehicles, buildings, and blimps as a means of heightening the public's awareness. As ratings became more important, stations became more cognizant of the need to promote.

Today, some call letters take on a special theme, promote a certain feeling, or tell a special story. Since radio is not a foreground medium, you need to be different to get the listeners' attention and leave an impression that will build awareness in your potential audience. Radio stations have created thematic identities such as, "The WAVE (KTVW)," "Power 106 (KPWR)," "WHUR-FM, Washington's Super Station," "K-Earth (KRTH)," "Coast Radio (KOST)," "Kiss (KIIS)," "K-Rock (KROQ)," and "K-Joy (KJOI)." Across the country, radio stations are coming up with all kinds of call letter identities, themes and logos (some funny, some not so funny). There are call letters like KFEL (Kan't Find Enough Liquor), WSB-(Welcome South, Brother), WMBD (World's Most Beautiful Drive), WSPD (Speed Petroleum), WBNS (Wolfe's Banks, Shoes, and Newspapers). And, there are many cocktail-hour creations including the Chicago myth offered a little over a decade ago by Harry Trigg: WLS (We Lose Sleep), WBBM (We Buy Bad Movies), WTTW (Wait Till This Weekend), WSNS (We Sell Nude Snapshots). It took rating services a while to adjust to this new world of sloganized identities because these services were not sure these stations delivered what their call-letters or slogans say they do.

Remember, in everything you do the basic thing you are promoting is your call letters and your dial position. Take every available opportunity to use your call letters and your frequency.

PUBLICITY AND SPECIAL EVENTS AS PROMOTION

Publicity refers to space in print publications or airtime on other broadcast stations for which the station makes no payment. Stations participate in parades and concerts and sponsor charitable fundraising activities in order to generate a positive public image. However, publicity is not guaranteed because it depends on the interests, needs, and discretion of the editors, managers, and owners of other media, who are mostly practitioners with their own agendas.

In seeking to gain publicity through other media, radio stations recognize that other stations rarely are interested in publicizing their activities. Why should they aid the competition? However, they may be interested if the information is of wide public interest.

When you are planning a special event or campaign, good timing can make the difference between publicity success and failure. Consider these suggestions:

- Take advantage of the fact that newspapers have more ads—so they need more editorial copy—before Christmas, Thanksgiving, and Mother's Day. Try for a seasonal or holiday story angle.
- Coordinate publicity timing with your employee's schedules. For example, before publicizing your telephone hotline, make sure your best hotline workers will be available when the news breaks.
- Call reporters and editors when they are not on deadline. Call morning newspapers in the morning, afternoon newspapers in the afternoon. Don't call broadcast stations right before they get on the air.
- For hot news events, consider media deadlines in scheduling the event. Give enough advance notice. Try to give competing publications or broadcast stations an even break; they will remember whether you did.

Placing Ads

Did you know that the placement of ad components can directly influence the consumer? This is because different placements appeal to different parts of the brain. This is how it works. Peripheral vision plays a big role. Much of the information consumers process about an ad comes from peripheral vision as they scan an ad or read editorial material next to it. While a person scans, the right field of vision is processed by the left side of the brain, which performs repetitive tasks and processes information presented sequentially. Therefore, copy components of an ad work better when placed to the right. The left field of vision, however, is processed by the right side of the brain, which is better at processing pictorial materials. Photos and other art should work better if placed to the left.

Pictures

In your promotion activities you will have occasion to take pictures for securing local, national, and trade publication publicity. Bear in mind that you will be sending these pictures to editors who receive hundreds each week, so the ones used must be of good quality. You might have a professional photographer take some decent pictures for you, but the topic, message, and details of that picture are still your responsibility.

The first thing to remember in setting up a publicity shot is that a good picture tells a story. The caption can give the detail of the story. But the story itself must be *in* the picture. The next thing to keep in mind is that you want a picture of *action*, not just people.

Always try to have your picture demonstrate graphically the story you are telling. Jack McDonald, in his *Handbook of Radio Publicity and Promotions*,[1] gives this example:

> A new disc jockey from Texas has been hired by your station. You have three choices of the type of publicity picture you can send out. (1) A straight head shot, or portrait picture, of the new man. (2) A picture of your station manager shaking hands with the disc jockey and welcoming him to the station. (3) You dress your new man up in a cowboy outfit complete with a Texas ten-gallon hat. You put him on a horse outside your studios. You place your station manager on a ladder next to the horse and then have the two men shake hands. Or better yet, have the Texan holding a six-shooter and the manager shaking that. If your were an editor, which picture would you use?

PUBLIC RELATIONS AS PROMOTION

The term *public relations* covers essentially everything that affects how people perceive your station. Perceptions may be affected by things as basic as the way a secretary answers the telephone, the condition of the visitors' bathrooms, the location and condition of your facilities, or by station involvement in a community fundraising activity.

To make this effort a success, all employees must understand the importance of this relationship and always act in ways that will reflect well for the station. Stations engage in planned activities to maintain a good public relationship. Such relationships include the following.

1. *Speakers Program (Bureau):* Station executives and other staff members give talks to classes in schools and colleges and to community organizations.

2. *Public Appearances:* Station employees, especially radio personalities, news anchors, and reporters, appear at events in the community such as charity fundraisers.

3. *Open House:* The public is invited to visit and tour the station and meet with members of the staff. The installation of a new piece of equipment, the station's anniversary, or an addition to the building are perfect occasions for an open house.

4. *Sponsorships*: Stations sponsor student scholarships and awards for people who have made significant contributions to the community. The station may sponsor sports events, children's art exhibitions, and concerts.

5. *Employee Participation:* Stations also encourage their employees to participate in organizations and clubs.

Public relations and publicity are often intertwined, even though it is a common practice to distinguish activities such as writing press releases and making speeches to organizations and groups as *public relations*. Radio stations issue press releases to announce a new format, a change in station management—new appointments, retirements—special programming, and new community activities sponsored by the station. The aim is to create an awareness in the minds of editors who might write related feature articles.

Writing a daily or weekly press release is usually a promotional activity of the station's marketing department. *Merchandising* is a public relations/publicity tool used by stations to remain popular with their listeners. A station gives away T-shirts, pens and markers, transistor radios with a fixed dial position of the station (Target Tuners), sun shields for auto windshields, coffee mugs, calendars, clocks, posters, bumper stickers, key chains, or a magazine, such as *The Country Beat* issued by WMZQ, a country-oriented station in Washington, D.C. The *WHUR Target Tuner* (Howard University Radio) is a pocket-sized walkman that fits into the palm. It displays the station's logo, address, and a listing of top programs. All of this leads to making sure the station's name is highly visible.

COPYRIGHTS AND RELEASES

Radio stations can register their slogans, theme statements, and logos as trademarks. This gives them protection within the United States. This form of protection can be given to attention-getters, characteristic sounds, call letters, and original program titles.

The Copyright Act of 1976 (effective in 1978) protects producers of broadcast scripts, photographs, slides, maps, and motion picture soundtracks. All music registered with Broadcast Music Incorporated (BMI), the American Society of Composers, Authors and Publishers (ASCAP), and the Society of European Stage Authors and Composers (SESAC) is protected and cannot be used without payment to these music-licensing organizations.

If you decide to use a photograph or drawing that was not created by your station personnel, make sure you secure permission to do so. Some of these photographs and drawings can be purchased outright or leased from the owners.

Release Forms

Another area of legality is that of using someone's face or voice in programs or promotional spots. Make sure the subject *signs* a release form permitting the use of his or her photograph or voice. This applies both to hired actors and to members of the public; it does not apply to news footage, celebrities, or public officials. Shots of people participating in large-scale public rallies, fairs, or parades are exempted from the need for releases.

In cases in which the photograph or voice of a contest winner is to be used to promote that campaign, a release form giving you permission to use the winner's face or voice must be signed by that contest winner.

Unless a member of the public seeks participation in your recording, a signed release must be obtained. When in doubt check with your station's legal counsel, the Screen Actors Guild, (SAG), or the American Federation of Television and Radio Artists (AFTRA).

NOTES

1. Mcdonald, Jack. *Handbook of Radio Publicity and Promotions.* Blue Ridge Summit TAB Books, Summit, PA, p. 17.

5
Planning the Promotion Campaign

OBJECTIVES AND GOALS

The promotion mix is usually coordinated on a *campaign* basis, making the campaign the relevant unit of promotion strategy. A campaign may last a few weeks, months, or the year; or if successful it may run for several years (for example, Avis' "We try harder" theme, or "Coke is the real thing" or the "Pepsi generation"). The most desirable marketing or promotion effort includes a total campaign with one unified theme. All promotional messages tie into this theme in one way or another, rather than conflicting with it.

A promotion campaign represents a consistent effort to implant the image of your station in the listener's mind. Using various positioning strategies, it tries to

- build program and station popularity
- develop station long-term loyalty that results in extended listening
- identify with the needs of the community (the coverage area)

The broadcaster's objectives are always to keep the tuned-in audience listening and get them to come back more often and for longer periods.

Broadcast marketing's first task is to define the audience a station wants to attract. Commercial broadcasters are strongly influenced by their advertisers' requirements and tend to define their audiences as those within the sponsors' market.

When setting promotional goals, commercial broadcast *marketing managers* (MMs) or directors (MDs) deal with an uncertain commitment from their management. In the ideal situation, it is the MD who each year guides the department heads and station management toward agreement on a set of goals for the coming year.

Broadcasters have just begun to face the need for 3-year and 5-year marketing plans, although stations have developed long-range budget and capital equipment projections. Long-term promotional plans create measurable criteria. They provide standards for comparison from year to year. These long-term plans also give a sense of direction to everyone involved, a direction that tends to build confidence, morale, and momentum.

In broad terms, the main objective of any radio promotion is to shift, establish, or solidify market perceptions. In other words, a radio station engages in promotion

1. to solidify its leadership position despite aging on-air personalities and audience demographics
2. to counteract a challenger who may be engaged in building and communicating stability, aggressiveness, and identity in the market
3. to turn itself around from a loser's position
4. to re-establish itself in a market if it has radically altered its programming or format.

First, it is advisable to write an objective statement (of 25 words or less). This statement should focus on the intended *results* of the campaign and not on the *process* for achieving those objectives. Objective statements that specify process, such as, "to show off new anchor people to the community," provide very little guidance or support for the creative process. Statements that focus on results, such as, "to establish the station as the leader in community activities" or, "to create an exciting, alive presence in the community," can be valuable in determining what decisions must be made to enhance the production process (e.g., press coverage, photographs, etc.).

A good radio promotion—whether sales promotion, audience promotion, or station promotion—aims at overcoming listener ignorance about the product or service by providing information. A good radio promotion also aims at overcoming listener inertia by using persuasion to create a favorable psychological association. This may induce the listener to act as the promoter wishes.

In formulating a promotional strategy, the PD must be guided by two major concerns: (1) projecting an image that would appeal to the target market, and (2) integrating the promotional efforts into a carefully designed concept geared to consumer preferences.

Promotion Planning: The Birth of Ideas

Good promotion takes a lot of time and hard work. It takes careful planning and proper execution. Like a gymnast who executes a flawless routine, making a promotion look easy is part of the process.

When you are trying to create an interesting and effective promotion campaign with a limited budget, fresh ideas become very important. Remember, promos for local stations have *very local* budgets. But redemption comes from the fact that the more interesting the promotion, the less you have to pay to keep the attention of the audience.

Trying harder is rarely the path to outstanding success. Trying smart with good ideas is the better way. Good ideas come from a variety of sources and *brainstorming* is a perfect outlet for generating promotional ideas. Done well, it is communication at its best—people feed off each other's ideas, random thoughts are recorded and all of the ifs and buts are explored.

During a brainstorming session for a promotion campaign, do not pass judgment, criticize, or evaluate ideas. People feel freer to share ideas when they know they are not going to be laughed at or criticized. Remember, no idea is too dumb or farfetched. Brainstorming is an arena for those "it'll never work" kind of thoughts. The point is to get many ideas, from which a quality concept can be selected.

Since no judgments are allowed during brainstorming, if you think an idea won't work, offer a variation that might work, and a proposed plan to make it feasible. It is also important to write everything down. The outcome of this session should be a long list of ideas from which to choose. Remember, there is no corporate structure in a brainstorming session. All participants have equal say, equal time and equal credibility.

Planning and Projected Costs

Projected costs are an important part of promotion planning. A station's promotions budget may depend on the size of the station, promotion mix, and level of competition.

Some of the costs involved can be met by a number of trade arrangements with companies that provide items such as T-shirts, posters, buttons, bumper stickers, billboards, and newspaper and television advertising. One method is to barter airtime for goods and services. Since PDs frequently negotiate trade arrangements with merchants as a way of defraying costs, it is essential that the PD understand the station's rate structure. Most stations prefer to exchange available airtime rather than cash for goods and services needed in a promotion.

The planning and implementation of certain promotions may require the involvement of consultants who possess the expertise to make sure your promotions are successful. Contests can turn into nightmares if improperly or unprofessionally handled. When in doubt ask the experts. Rick Sklar, onetime program director for WABC, New York, is credited with running some of the most successful radio promotions ever. In his autobiographical book, *Rocking America*,[1] he tells of the time that he was forced to hire, at great expense, sixty office temporaries for a period of one month to count the more than 170 million ballots received in response to the station's "Principal of the Year" contest. For a similar contest the year before, WABC received six million ballots.

Writing about proper planning and the use of consultants, Mr. Sklar tells about another occasion when he had more than 4 million WABC buttons made as part of a promotion. The station awarded a cash prize to listeners spotted wearing them. Unfortunately, he did not factor in the huge cost involved in shipping millions of buttons from different parts of the country. The station had to come up with thousands of dollars to cover air freight. Proper planning during the developmental phase will prevent unpleasant surprises—especially unbudgeted expenditures.

PROMOTION AND THE STRATEGIC MARKETING PLAN

While promotions are often the most visible face of a radio station, they have to be more than just an image or stunt. They have to be in concert with the station's marketing goals and objectives. Promotions must become an integral part of the marketing process rather than an unconnected special event.

Great promotions always begin with sound thinking and good ideas. While innovation and creativity are key elements in all good promotions, proper planning and execution are just as important.

Any evaluation of a promotion campaign should revolve around the station's *strategic marketing plan,* simply known as SMP. Properly developed, the SMP is a structure—a way of keeping track of your station, your competition and your progress. It is not something you prepare at budget time, then toss in a bottom drawer and forget.

A good way to begin is to suggest to the GM that—if your station does not already have an SMP—you would like to work with the GM and others in taking the leadership in preparing such a plan—with full staff input and support. Have a rational plan with clearly defined goals and objectives.

Appeal to the GM's (or owners') self-interest by stressing the benefits that will accrue from the ideas that come out of your brainstorming session(s). Back your presentation with solid, reliable information or evidence; prove your case, get the GM, the program manager (PRM), the sales manager (SM), and news and community affairs director (NCD) involved. Earn their respect and trust by listening to them and by talking openly and honestly.

The following suggestions might help you sell your ideas, get management support for those ideas, and create the team concept that you need to be a successful PD:

- Keep the lines of communication open. Let others know what you think and learn to draw out their thoughts.
- Share your problems. Let your supervisor know when things are not going well.
- Don't shy away from confrontation; it can be constructive. As part of the confrontation process, look for ways to win others to your point of view.
- Learn to think several steps ahead, so that you can cover all your bases.
- Generate excitement, interest, and curiosity by presenting your idea creatively. For example, present a tease of your idea first—tell a little now, more later.
- Get feedback from coworkers.
- Finally, choose the role, if appropriate, of cheerleader, networker, or politician at your station. Show your enthusiasm. It's catching.

If you have trouble delegating work because you feel you can do the job better and faster, think of yourself as a coach rather than a manager. The coach trains, motivates, and encourages others to do their best but never touches the ball during the game. Good supervision is the art of getting average people to do superior work.

The Promotion Director: Interrelationships

Your SMP must be flexible and readily adaptable to a changing environment—the ever-changing world of radio broadcasting. In ideal situations, an SMP is developed by the PD or the MM, along with active input and support from those people who help to make marketing decisions that effect the station's direction. These will naturally include the GM, PRM, and the SM. In situations where these positions exist, the ND and/or a CRD might be involved. With this kind of cooperation and backing, you are ready to prepare your outline.

Step One: Analyze Your Station

Begin by listing your station's key strengths and benefits. Be honest and thorough. Be realistic. Are you really the "sound of the city" or "the hit music, free money" station? Do you really play "more music all the time"? Are you really the "news leader?"

BPME's *Image* suggests that you sincerely try to answer the following questions:

- How effectively does your station implement its format?
- What benefits does your station provide its audience?
- How does your station and its staff positively interact with and support the community? Be painfully honest.

Concentrate on strengths that your listeners recognize about your station—strengths on which you can capitalize to achieve your goals. Now list your station's key weaknesses or drawbacks. Consider the following:

- Do you play more commercials than your competition?
- Are your contests really that exciting?
- Is the audience really interested in the topics of your talk show?
- Are your personalities strong?

To help you formulate this list, think as if you were the competition. What could they promote against your station? Again, be honest and objective.

Step Two: Analyze the Competition

First, list all competing stations in your market. Put each on a separate sheet or use a word processor for easier arrangement. Determine the position of each station. What does it represent in the mind of the listener? What images? What symbols? How does the listener perceive the station? Now, rank all the music stations in the market—from hard rock to classical. This might reveal some new competition. Concentrate only on those stations that you feel are competing for a similar audience. For each station, perform a strength versus weakness analysis similar to the one you performed in Step One. Look carefully for opportunities to exploit the weaknesses of the competition and be alert for potential threats, because you must assume they also are analyzing your station in the same way in their SMPs. This exercise is commonly called a SWOT analysis (strengths, weaknesses, opportunities, threats). Some PDs refer to their planning group as a S.W.O.T. Team.

Step Three: Analyze the Audience

Start by identifying your station's audience. Use arbitron data, programmers packages, or any information you can obtain, and develop a geographic, demographic, and psychographic analysis of your audience. Your research should tell you the following about your audience(s)—who they really are:

- age group
- predominant sex
- ethnic characteristics

- income range
- neighborhood(s) where they live, work, and play
- recreational activities
- educational levels

Attempt to understand why listeners tune to your station and what benefits your station—and the competing stations—provide for them.

Keep in mind that this is not an exact science. However, the process of learning about the audience will help in future marketing decisions and sales efforts. As a part of this process, perform a similar analysis of your competition.

Step Four: Analyze and Set Goals

At this point, you have developed a long list of strengths and weaknesses for your station and your competitors. You should have a feel for the composition of the audience in terms of their geography and lifestyles. This information will help you create a *planning framework*.

You now are ready to determine which of your strengths are strong enough to overcome your weaknesses.

- Which of your station's weaknesses will you shore up, how should you shore up these weaknesses, and how will you accomplish this task?
- How can you turn your competitor's strengths into weaknesses and then exploit them?

In this process you will be developing a preliminary set of goals. By referring to the audience analysis you will be able to refine these goals and establish a priority list. The following are some tough questions to test your objective statements:

1. So what? Who cares? Is it clear *why* you are running this campaign at all?
2. What's in it for your audience? Is there a specific listener benefit?
3. What results (short- and long-term) do you expect from the campaign?
4. How do you want the listener to feel about your station after the campaign? And how is that feeling different from the way they feel now?
5. Is impact important to achieving your objective? If it is, does the statement reflect that?
6. Is it doable? Do you have the staff, time, and budget?
7. Is the objective statement specific to the strength of your station and your current position in the market?
8. What makes your campaign different from those of your competitors?
9. Is everyone involved willing to actively support the achievement of the objectives?

The statement of objectives should be complete, accurate, and doable within the time frame. Once you get everyone to agree with the objectives, treat it like a contract. The statement of objectives should stay with the campaign from concept to completion, from brainstorming, writing, and design, to production, and postproduction. (Be sure to communicate the purpose and objectives of the campaign with the research, creative services, music, and graphic design departments.) These groups could become powerful support tools for you. Ask yourself these questions as you build your promotional plan.

How will your station benefit?

1. Will the promotion appeal to your target audience and increase listenership?
2. Is it unique and creative enough to distinguish your station from others in the market?
3. Does it provide opportunities for listeners or prospective listeners to name your air product or identify with your station?
4. Are there legal uncertainties about the promotion?
5. Is it timely?
6. Is it targeted to the appropriate demographic/psychographic group(s)?

How will your client(s) benefit?

1. Will the promotion attract new consumers for your advertisers?
2. Will local clients enjoy their participation or will it be an extra hassle for them?
3. Is the relationship between the promotion and the client's business goals clearly communicated to participating advertisers?
4. Will it attract new advertisers for the station?
5. Will it give current advertisers added value for their investment?
6. Can the promotion be used to extend or expand existing advertising commitments?

Can you get press coverage?

1. Do you have personal contacts and up-to-date mailing lists ready to alert local newspapers, television stations, trade, and nontrade magazines?
2. Are you sending press releases and making phone calls?
3. Are you thinking about creative ways to get the press excited about your promotion?
4. Have you invited reporters to visit the station or attend the special event?
5. Is there a public service angle that merits special attention?

The idea behind any promotion is to win listeners. Over the years stations have used a variety of methods to reach this goal. To gain the listening public's attention, a California DJ set a world record by sitting in every seat of a major league ballpark that held 65,000 spectators. In the process, the deejay injured his leg. However, he accomplished his goal of garnering national attention for himself and his station.

One of the most infamous examples of a promotion gone bad occurred when a station decided to air-drop dozens of turkeys to a waiting crowd of listeners in a local shopping center parking lot. Unfortunately, the station discovered too late that turkeys are not adept at flying at heights above thirty feet. Consequently, several cars were damaged and witnesses traumatized as turkeys plunged to the ground. This promotion-turned-nightmare was depicted in an episode of the television sitcom "WKRP in Cincinnati."

The list of glitches seems endless. A station in California came close to disaster when a promotion that challenged listeners to find a buried treasure resulted in half of the community being dug up by overzealous contestants. The promotions did catch the attention of the public, but in each case the station's image was somewhat tarnished. The axiom that any publicity, good or bad, is better than none at all can hurt a station, contends Chuck Davis, PD WSUB/WQGN, New London, Connecticut. "It's great to get lots of exposure for a station, but if it makes the station look foolish, it can work against you." [2]

Most radio promotions are practical. They offer big prizes rather than stunts, and they tend to draw the most interest. In the mid-1980s, WASH-AM, Washington, D.C., and KSSK-AM, Honolulu, both gave a lucky listener a million dollars. Cash prizes are a big draw. Valuable prizes other than cash also can boost ratings.

Step Five: Develop Goals and Priorities

Using all the information you have developed, determine where you can act. Choose your battle carefully in light of limited resources, but act decisively to reach your goals. Should you use billboards in the suburbs? Is late night television a valuable component? Should you become involved in contests? How many? When? Should you become involved in community events? Which ones? Should you have money giveaways? How much and when? Remember to build as many client tie-ins to as many promotions, activities, and events as possible.

As you set priorities, you will reap the benefits of having a good analysis of your station's market, competition, and audience. Determine exactly who you want to reach with what message(s), what the best ways are to reach them, and what is the best timing. Be sure your goals are consistent with your strengths and weaknesses and with those of your competition. Develop your theme(s). Be creative.

Step Six: Evaluate and Update

One of the strong points about this plan is its *flexibility*: it can be continually changed. Continually evaluate your successes and failures. Monitor your market situation as the days and months go by and note changes, such as other station format changes, ad campaigns, contest activities, and rating shifts. You will be in a position to counter, adjust, or initiate new actions from a position of knowledge and strength. Be positive, and set a positive attitude for all at your station regarding its position and the various elements of your SMP.

In the competitive promotions environment, advertisers are spending millions to be more memorable, to be more provocative, and to get more awareness and response. Therefore, every person involved in the creative aspect of promotion must do all they can to keep their message fresh. They have to strive to be unique. This starts with an understanding of the market and the promotions business, and a familiarity with the goals of the station and the campaign. It means working with people who share your vision and commitment.

Jack McDonald, in *The Handbook of Radio Publicity and Promotion*, suggests that you associate your promotions with community events. If your promotions are logically associated with the things that are happening in your community, he says, "they will have added impetus, the extra support to make them real winners." [3]

Before getting involved in a sensitive, political community issue, be sure to consider these points. Whether the issue is gun control or saving a community landmark for demolition, it is imperative to have a solid understanding of your audience's viewpoint. If there is the slightest doubt whether you will receive their full support, rethink your position.

Chuck Davis, PD WSUB/WQGN, New London, Connecticut advises that "it's great to get lots of exposure for a station, but if it makes the station look foolish, it can work against you." [4]

Performance Standards and the Evaluation Process

Close monitoring of progress means breaking down broad standards of performance to specifics such as ratings, client roster, and airtime sales goals. Defining goals is a vital task but not always an easy one. Working closely with the sales and marketing department, the PD has to consider such elements as past sales performance, new accounts, past and present sales research, and expectations of top management.

After receiving this information, the PD and staff will have to figure out how much of it should be put in the nice to know category and how much is conclusive evidence that a specific promotion campaign brought added listeners, which translated into more advertisers and more revenue for the station. Procedures for *evaluation* should be set up before they are actually needed. Management by crisis doesn't always work. Of course, it is impossible to make every decision in advance because the unexpected is always happening. But a good PD or MM will schedule points of decision at the beginning of a campaign.

Assuming that it was a good campaign, the PD will be able to study the results and plan for the future. Suppose the campaign was unsuccessful, your station now has a chance to take *corrective action* to avoid whatever problems the failed campaign might have encountered. In doing this, you should determine what the current situation is, evaluate possible modifications of existing strategy, and, if necessary, come up with a whole new strategy.

Promotion decision makers face three procedural problems in taking corrective action: identifying the causes of the problem; adjusting for time lags between the point of decision and the point of action; and budgeting elaborate and costly control procedures. Costs may be so heavy that they outweigh benefits.

NOTES

1. Sklar, Rick. *Rocking America*. New York: St. Martin's Press, 1984, p. 52.
2. Keith, Michael C., and Joseph M. Krause. *The Radio Station* (2nd Edition). Boston: Focal Press, 1989, p. 164.
3. McDonald, Jack. *Handbook of Radio Publicity and Promotion*. Blue Ridge Summit, PA, pp. 13–18.
4. Keith, Michael C., and Joseph M. Krause. *The Radio Station* (2nd Edition). Boston: Focal Press, 1989, p. 162.

6
Promotions That Work

The radio promotion professional's job is not an easy one, and it's getting harder all the time. With the abundance of format choices available to listeners today, we need to sharpen our skills to the finest degree to cut through the promotional clutter and reach audiences with bold, innovative, and powerful messages to sell products. Whether creating a print ad, radio spot, or on-air promotion, your selling strategies must be precise and clear-cut. No amount of audio or state-of-the-art technology will sell your program. As a matter of fact, these clever and often entertaining pieces sometimes serve only to obscure the primary marketing message. The only promotions that mean anything are the promotions that work, sell, and earn an audience. The more effective your sales message, the more powerful your promotion. Judge your promotion by asking the following questions:

- Is the promotion simple?
- Will it take a lot of time?
- Can it be explained easily and quickly?
- Can anyone in your listening area participate?
- Is there an element of suspense?
- Is it necessary to be listening to your station in order to play the game or contest?

If at least the majority of these elements are not present, reevaluate your promotion and find a way to include more of the points in those questions.

It might be unwise to look at every promotion and think it's perfectly translatable to your station. You might need to customize it to fit your particular needs. When you are planning a promotion, look at the format of your radio station, look at what your customers want from you, and then make your decisions. Combine the power of a good creative idea with the power of radio.

GETTING YOUR CLIENTS INVOLVED

Get your clients involved. When the promotion idea was introduced to your clients, did their eyes light up? What will this do for the client's business? Will it create traffic for the client? What are its revenue-generating potentials? Will such a promotion give your station some visibility?

Advertisers have long demonstrated an active interest in on-air promotional activity, whether their product or service is featured in paid commercial announcements or promotional inclusions. Some years ago it was revealed that advertisers would pay for promotional inclusion over and above commercial announcements. The payment by advertisers usually amounted to "trade for mentions" but that has evolved to the point that now most leading advertisers will pay money to stations for promotional involvement, no commercials necessary. The value of being featured in a station's promotional contest announcements adds weight to their advertising and merchandising efforts over and above commercial impressions. Exposure of an advertiser's product or service to a core audience within the umbrella of a station's call letters provides them the *reach* they cannot achieve in commercial schedules.

Advertisers are particularly interested in promotional concepts that provide in-store traffic or on-air product exclusives. They are looking for opportunities that provide the inside track to a station's or program's loyal audience core. Advertisers are most interested in promotions that make both their customers and the station's audience win simultaneously.

An example of this principle is a radio promotion called "The Magic Ticket." The concept is simple: a station's listeners are offered chances to win hundreds of thousands of dollars in prizes and cash by going into a retailer's stores—with multiple local outlets—to pick up individually numbered tickets, no purchase necessary. The station can pre-announce the times when ticket numbers will be called out. This contest is exciting for players and listeners, the advertiser pays for everything, and the ratings will soar. This promotion has appeared in more than 50 markets.

The difficulty many stations have with client on-air promotional contests is designing effective marketing plans that accommodate both their programming needs and the client's merchandising objectives. Many stations wait for an advertiser to come to them with their promotional needs. The downside of that approach is that advertisers frequently pit one station against another to get the best airtime deal. According to BPME's *Image*, many stations are reluctant to design aggressive promotional concepts that involve advertisers on a paying basis for fear of being turned down and losing face. Some stations, it continues, lack confidence in their ability to execute a tightly coordinated promotional contest involving paying advertisers for fear it may not come off as promised and may, as a result, damage the relationship with those clients.

Ask anyone who has to be creative for a living and he will agree that good ideas often evolve out of bad ideas. Sometimes crazy thoughts will spur practical ones. Perfectionists must relax their standards and recognize that brilliant ideas often begin life as far-from-perfect thoughts.

Here is a little exercise that will get you going in the right direction. Try to think up 15 uncommon uses for a common straight pin. Start with five practical uses (e.g., a bulletin board pin, a crevice cleaner, an instrument to draw boils), then move on to list five over-the-rainbow ideas (e.g., a sword for an elf, an antenna for the world's smallest radio station—or an idea as fanciful as something to provide the meaning of life to a pincushion). After you have pushed your mind to silly extremes, return quickly to more practical concepts for your last five ideas. John Emmerling, in his book, *It Only Takes One: How to Create the Right Idea*,[1] suggests that you may find that the last five ideas are the best, solid ideas equipped with a little pizzazz.

SELLING PROMOTION IDEAS TO YOUR CLIENTS

Make sure you present all promotional ideas to clients in as sophisticated and professional a manner as possible. Use an artist who has handled all your business and will therefore know your station's style. Graphics can range from posters and audiovisual materials to mock-ups of point-of-purchase displays or posters. Contest ideas are as follows:

- Give away trips to Florida (or anywhere warm) if you are in a cold state.
- Celebrate anniversaries: give away a vintage car to celebrate the station's 20th, 30th, 40th, or 50th year. Some of these cars cost between $3000–$5000, which could be half cash and half trade for advertising.
- Help host an annual, old-fashioned Fourth of July. Bring in local and outside bands, refreshments, and have fireworks in the park. Include several clients as sponsors. Put up posters all over town before the event, featuring the sponsors.
- If in a rural town, help sponsor "Farmer Appreciation Day," local merchants can work with you to present daily specials, and you can present big posters for them.
- Give away a Volkswagen–"Win Your Own June Bug,"—in June. If you are smart, you can have an artist paint your logo on the vehicle. If you are lucky, the winner will leave it that way.
- Give away a zero coupon bond that will be worth $10,000 in 20 years. It only costs $600–$700 now—but be sure to review your state's laws concerning this giveaway.
- Give local stores, high schools, and colleges free poster boards—they can use them for anything they want. Naturally, your call letters, dial position, and logo are at the bottom.
- Send a newsletter to clients, incorporating some of the materials sent to your station from sources such as the Radio Advertising Bureau (RAB), and include quick write-ups of recent promotions and how they worked for clients. Happy letters or success letters from clients can also be included. Check with anybody whose material you intend to republish to make sure it is OK to do so.
- Offer to provide grocery and other retail stores with small shelf signs that say as advertised on your call letters.

CREATIVE PROMOTIONS

The following are examples of creative promotions that have been tried and tested. There are numerous types of promotions, and this is a brief look at the essential ingredients of the top ten of the most commonly mentioned types of promotions and promotional ideas.

Traffic Builders

KRAK, Sacramento, California, did a National Football League (NFL) schedule promotion. In conjunction with the NFL, KRAK printed about 25,000 schedules each week during football season. Inside each schedule was a serialized number. Each week during broadcast, the station drew a number to match up against its mailing—(schedules that had been mailed to listeners). The schedules were distributed through retail outlets, thus creating massive traffic. The lucky winner with the serialized number of the week won a trip for two to the Superbowl. This was done for the entire 17-week football season. Also inside the schedules were six different coupons for advertisers. KRAK was not only selling the trafficking aspect of the promotion, they were also selling the coupons themselves.

Scan and win is a good example of a traffic builder. Have your station print promotional materials with bar-codes. Only a certain number of bar-codes will trigger the scanner to say it is a winner. Bar-codes could be printed on cards that are mailed out by the station or a bank, distributed during station remotes, put on car windshields or they could be an on-air contest. Listeners can win from two to ten bar-code cards. The idea is to get them out to people. The people with the bar-coded pieces go to various retailers (these can be different retailers every week or for periods of six weeks or longer) to see if they've won something. The retailer runs these bar-coded pieces quickly through the scanner. Gold Media Corporation has developed a scanner for contests of this type that says "congratulations" for winning, or suggests that you "try again: better luck next time" if you haven't. These scanners can be leased by your radio station. Introduce this to the retailers and they'll understand quite quickly how it will create traffic. It can be used to get people into shopping malls, car-dealerships, or any retail store.

Sponsor Tie-ins

These are tie-ins to listener events. An example of this is an off-shoot of the office party. A station holds a large office party two or three times a year. The sales goal is to get an ongoing relationship with the hotel chain(s) in which the station hosts these parties. Have joint sponsorship with the hotel as one of the sponsors. These office parties are fun events and could be billed as "The World's Largest Office Party." There is value added for the client that goes beyond just buying a spot schedule. Those people want a relationship with your station because you have something they could use: listeners, which translates into potential customers and revenue.

The Plastic Card Promotion

Several years ago, broadcasters discovered the magic of combining the mystique and long life of plastic with the unrivaled savings appeal of coupons. A new and different kind of promotion, the advertiser-funded, contest-oriented, *plastic card* promotion appeared on the scene as an innovative way to boost ratings and sell more time (Figures 10 and 11).

Promotions That Work 55 ▼

▶ *Figure 10* Radio stations KHEY, Z104-FM and WGCI have run very successful "Plastic Card" campaigns. Used with permission.

WLS' "Fantastic Plastic" In the spring of 1980, WLS Radio startled its Chicago competition when it blitzed the market with 1,500,000 "Fantastic Plastic" cards, produced by 2B System Corporation, of Madison, Michigan. The majority of the cards were distributed through inserts in a Sunday edition of the *Chicago Tri-*

▶ *Figure 11* Supermarkets, restaurants, auto dealers, fast food chains, retailers, video and record stores, and local merchants and franchises of every description use the 2B Plastic Card Promotions. Used with permission.

bune. The remaining cards were given out to residents by participating retailers. The format was an $8\frac{1}{2} \times 11''$ carrier to which the plastic card was attached. The carrier described the multitude of benefits offered exclusively to cardholders. The plastic card was easily removable. An important part of the carrier was a clip-out activator card, which the recipient was required to fill in with name, address, card number, and other data, and mail back to the station in order to become eligible for cooperating merchant discounts and station prizes. This format with the activator card has since become standard with the hundreds of stations. However, carrier size sometimes varies.

WLS "Fantastic Plastic" was planned originally to be only a spring and summer promotion. But ratings went up consistently; advertisers, according to the station, actually vied with each other and waged vigorous campaigns to become sponsors, and the promotion remained a year-round concept. "Fantastic Plastic" became a byword in Chicago and this card remained with the station for years. Before long, plastic card promotions were catching on all over the country.

The fact that the plastic card promotion is advertiser-funded is one of the major appeals to GMs and PDs. Stations differ in whether they go with a major almost exclusive sponsor or with several almost equal participants.

One strategy is to combine a major sponsor with several minor ones. Space allocated on the card-affixed carrier and on-air time is in direct proportion to the size of the schedule the sponsors buy.

Soft drinks and fast-food restaurants are the largest group of sponsors, followed by car dealers and home builders. Other frequently tied-in advertisers are furniture and appliance stores, travel agencies, restaurants, video and record chains, supermarkets and convenience stores, and banks and savings and loans.

How It Works The cards are sequentially numbered. This numbering avoids problems with cash calls and contest awards. How many Sam Smiths are there? Which Sam Smith are you? There's only one cardholder for each preregistered number. Cardholders must show their cards to cooperating merchants to get the benefits.

The plastic card gives the radio station an effective way to actively merchandise its advertiser's goods and services. Every time listeners take out the card to participate in a station's promotion, they *see* the station's logo and dial position. The station is with the listener all the time . . . 52 weeks a year.

Examples of prizes include a free car wash; free movies; sporting events and concert discounts; bargains at furniture stores; two dinners for the price on one; cash calls for specific card numbers; money prizes. There is always something new. Find ways to get cards into the hands of people who haven't had them before.

Your local situation, types of sponsors, availability of media, and station policy on trades can all influence the ways your station gets cards into people's hands. Utilize your Sunday supplement, direct mail, and daily and weekly newspapers. KABC, Los Angeles, which has been doing plastic card promotions for years, has hundreds of thousands of its highly successful "Talkradio" cardholders's names in their computer. When KABC launched its monthly magazine, *Let's Talk*, 250,000 copies were mailed to "Talkradio" cardholders.

WGCI, Chicago, tied in with high schools by inserting its "Card of Gold" into yearbooks. Through trades with sponsors, several stations have featured such prizes as free groceries or gasoline for a year. Other versions of successful stretched-out prizes have been paying utility bills, house payments or cash payments for one year.

Whatever the prizes, big or small, almost all stations have supplemented these awards with *cash calls*. A few years ago, WMAL, Washington, D.C., used its "Club 63" plastic card to give away a home, several cars, numerous trips, and even diamonds. The station aggressively promoted its "Money Line" phone number. By calling this number the listener heard a prerecorded message that told where discounts and special deals were available for "Club 63" members that day. The message carried plugs for advertisers and shopping hints about the best buys.

Stations contemplating plastic card promotions should, suggests 2B System Corporation, Madison, Michigan,[2] carefully check their promotion's structure for five strong building blocks:

1. Make it easy for people to get and carry your plastic card. On launch day, make your announcement an event. Structure your market.

2. Give your plastic card promotion a uniqueness. With heads-up planning, you can—through trading—load your promotion with lots of listener benefits (exotic prizes, multitudinous giveaways, shopping savings galore) that no broadcast promotion in your market can possibly compete with. Wrap all that bounty into one irresistible package, give it a singular, distinctive, easy-to-remember name, and promote it.

3. Stand on your own feet. Make your plastic card promotion 100% yours. Do not dilute your image or your impact. You have something no other couponer has: your valuable plastic card. Don't let your sponsor overshadow you. Be a star not a supporting player.

4. Be sure to incorporate the "Activator Snap-Out Postcard" as part of your promotional carrier. Remember, the person receiving your numbered plastic card fills in an activator card, affixes postage, and mails the card to you, immediately becoming a participant in your promotion. No longer a passive onlooker, this person has a vested interest in your promotion. This also gives you valuable demographic data and a mailing list for follow-up promotions—an actual count of people using your card and listening to your station. Your sales staff will find this very helpful.

5. Plan your plastic card promotion not as a one-shot deal, but as the basis for an extended campaign. This card can be converted into a permanent audience-boosting, revenue-producing vehicle.

On-Air Services

These don't necessarily have sponsor tie-ins that create traffic or an event. But there is something in the air that is good for the sponsor and the radio station. The on-air *restaurant guide* is done as a service for listeners looking for good restaurants. Credit card companies like Visa and American Express might find this very profitable. In fact, these companies could introduce your station to new restaurants in your market. Because it might not make sense for all of these restaurants to buy spots on the station, American Express or Visa could become an *umbrella* sponsor. In this case, three or four restaurants could be mentioned in one spot. Grouping similar services like this might not be offensive to listeners since they all have different cuisines.

Client Incentives

This involves doing things that will make your client spend more money on your radio station. You immediately think of incentive trips—and for a lot of people, they have been very successful. But one of the more original ones is one called an *action auction*. This is usually done during the first quarter of the middle of summer—the period that might be a little slow and need a boost. For every dollar that the client contracts to spend or does spend, your station gives then an *auction dollar*. Don't let your clients know what these auction dollars are for. These bills should be in bright colors and three or four times larger than actual paper money. Tell your advertisers that these dollar bills have a value and should be saved. Tell them you'll give them one of these large bills for every dollar they spend on your station.

When the promotion comes to an end and you've given away all the dollars, gather all the clients together (all those with auction dollars) at a nice hotel room with a partition. Serve some hors d'oeuvres and wine. Your clients are now in a room divided by a partition enjoying themselves, getting to know each other, and trying to guess what this is all about. They do not see the other side of the room. After they have wined and dined for about an hour, open up the partition to reveal a stage that has been set up for an auction. All the prizes are displayed. The first prize might be a fishing boat, maybe a computer, or a trip to Hawaii, and a lot of little prizes. It's up to you and your imagination. Then start the auction. Start with the small prizes, so that everybody is getting something for their auction dollars. The bidding will start to get a little crazy and by the time you get to the fishing boat, people will start pulling their auction dollars together for joint purchases. It has become such a fun event in some markets that clients look forward to this each year. And before it takes place, clients make deals and plans to group their auction bucks together to win big at the auction.

Station Promotions to the Client

A lot of stations get involved in the day-to-day promotions *for* their clients and don't do enough promoting *to* their clients. With all the competition in radio today, your clients are a group that should not be forgotten. Your station's image with clients could help keep your revenue steady when your ratings are not so hot. It is therefore important to meet with your SM and the sales team to find out what tools they might need to image themselves and your station to the client. What are they saying about your station? The station personnel? Too young? Too old? Too male? Too female? Too downscale? Too upscale? You can then devise some promotional materials to answer those perceptions and highlight your strengths. A few years ago, KGON, Portland, Oregon, did a profile brochure. For years, this rock station had been battling a heavy metal image, which was sometimes negatively interpreted by clients. The profile brochures directly addressed those perceptions and used research-based information to show that listeners aged 18–34 were actually buying more cars, more furniture, more groceries more than people in other age groups—all the things the station's clients were interested in. The key was to present this information in a factual manner, and to speak a language that sales people and their clients wanted to hear. The station ordered about an 18-month supply. The research paid for itself in two months with orders from clients who previously were reluctant to buy spots on the station because they felt only 13-year-olds in leather jackets without any money listened to the station. You don't have to do an entire brochure—one sheet that is factual and impressive might be all you need.

You might need to convince your clients that people listen to your station at work. Do an on-air contest and have listeners at work respond to you on their official letterheads. Make a copy of their entries and mail them to your client. There's no better proof that people listen to you at work. This could help sell your 9 to 5 package. A question that you should ask is who are the people buying the advertising? Have your sales staff put a bio together on the major clients, so that you know who you are marketing to. In a market where 75% of the major American buyers are females aged 25–35, if you're a male rock station targeting ages 18–34 or an easy listening or an oldie station, these key buyers might not be tuning in to your station. So what do you do to stay competitive if your numbers are not there? You need to develop materials aimed at presenting your station in a manner that buyers will understand, accept, and to which they can relate. Try to convince your buyers to listen to your format even if it might be a format that they say they don't want to listen to. Just let them know what you are doing and use ways to get them to try your station even if it's only for a day.

KJI in Seattle, ran an on-air contest a few years ago called "Sing It and Win," inviting listeners to call in to a special phone machine and sing its jingle (which had been around Seattle for years). If their version was played back on the air they won cash. Simple promotion. But instead of just telling their listeners about it, the station set up another phone line and answering machine for local and national sales and sent fliers asking clients to call in and sing and win. They heard a brief message from KJI's sales manager on the tape, and they then sang the jingle. Every day the tape was nearly full. Every time clients called in, they were given a $95 cash prize, and at the end of the promotion a thank-you card and a T-shirt were sent to all the participants.

It doesn't have to be a big prize. Promoting to your client shouldn't be short-term or just contest-based. It should be long-term and continuous. Alternate the contests with research-based information just to keep clients abreast of what the station is doing. Always focus on what you want to accomplish. Whether you use a sales piece, a poster, or a brochure, position your station as special. A few years ago, WBOS, Boston, launched a campaign for clients when its owners decided to change the station's format from country to album classics. Boston, like most major cities, has some very good radio stations competing for the advertising dollar. Advertisers are usually skeptical about new formats or management. They develop a "wait and see" attitude. They wait to see what the numbers (ratings) tell. WBOS's goal was to convince potential advertisers that it was a major player in the market with a long-term commitment to both the station (with its brand new staff) and its listeners.

The station planned a media information campaign aimed at presenting potential advertisers with attractive and informative pieces that would help image the station immediately. It included a multicolored mini-poster titled "We've hung up our spurs." Delivered with the poster was an invitation to listen to the station the day the format was changed. That was followed by a brand new sales kit and presentation folder. In a case like this, make sure your folder is different. Take a look at folders of other stations in your market. Try to make yours as unique as possible. WBOS made its folder horizontal instead of vertical. It included some detachable Roladex® cards with the telephone number of a 24-hour sales line. These popped out of the folder into Roladex files immediately. In short, look at what your station needs to accomplish; what message or image needs to be presented to your client; and how you can involve clients with your station. Use research to legitimize your points (demographics, market profiles, audience profiles). Know your clients and your competition. Who are you marketing to and what are the others stations in the market doing? Don't be the fourth station in your market to give a trip to Barbados—winners won't remember who they got it from.

Another bonus to marketing to clients, besides helping your sales, is increasing your recognition and worth to your GM. There's nothing better for catching a GM's attention than compliments or orders from a client for pieces you've created. You'll be showing your understanding of audience programming and sales, will shed the image of a party planner, which will help you develop to a full-service PD or MM.

On-Air Features

An inherent problem for news radio stations is how to do exciting things. You don't do contests, and you don't do major giveaways. There are no real "personalities." You cannot take people on trips. So what do you do to excite listener interest on an emotional level? KFWB, Los Angeles, came up with an idea.

KFWB is "Group W's All News Radio Station in Los Angeles." It is a constant cumulative ratings leader, a constant billing leader, and is always locked in battle with its competitor, CBS-owned KNX radio. KFBW is fraught with signal problems but has been around a long time and has a strong history for good solid news reporting. KFWB came up with a promotion it called "California Dreaming."

In February 1987, they took an entire 12-hour day, got rid of every commercial spot, and inserted either a listener, a movie star, a sports personality, or a corporate executive talking about their dreams for a better California. The station hired a production company to get people from all over the country in Los Angeles for a grand party and to make a short radio statement about their dreams for California. People called in, there was limousine service, and competitors (in business and politics) met for the first time. Listeners heard unknown people and people they thought too busy talk about something that meant so much to them. These statements were interspersed with pretaped suggestions from Los Angeles listeners taken on an 800 line.

This promotion did an incredible PR job. Every newspaper and TV station covered it. KFWB simultaneously broadcast the event on KABC radio. Hundreds of people participated and over a million listeners were reached by it.

The Great Grocery Giveaway

The objectives of this giveaway are to convert a regular newspaper advertiser into a radio client, offer listeners an adult-oriented incentive to tune in, build store traffic and demonstrate the effectiveness of radio ads to a major regional merchant, and link the excitement of the radio station to the store with a remote broadcast of the final contest round.

How It Worked at WELI (New Haven) The radio station worked with three local Pathmark grocery stores to create an awareness promotion that culminated in a supermarket sweep, which was broadcast as a live remote. 200,000 baseball cards bearing the likeness of WELI's on-air staff were distributed (one-per-customer) exclusively at Pathmark. Posters in front windows and at checkout counters carried WELI's logo and dial position during the course of the six-week ad schedule. Five times each day, Monday through Friday, a designated caller, holding the proper card, won a $10 gift certificate to any one of the stores. During the final week four finalists won $100 gift certificates for groceries.

The Investment WELI spent $2500 for the baseball cards (more than 150,000 were given out) and $75 for poster art. 60-second spots for the promotion were produced and aired ten times daily (Monday through Sunday) for six weeks. 15 ten-second "liners" (coming up contest, past winner names, designated callers, and other bumper-teasers) were randomly presented in all dayparts during each day of the six-week promotion.

Pathmark purchased a $10,000 advertising schedule and donated another $12,000 worth of groceries. They paid production costs for in-store window and checkout posters and included the WELI logo in six weekly newspaper inserts.

Seasonal Promotions

Weekend Escape There are year-round resort motels within an easy drive of almost everywhere these days. A good and inexpensive wintertime prize (resort motels will trade) is a weekend for two (or maybe with the kids) at a motel featuring an indoor pool, spa, court games, etc. Tie it in with an advertiser or group of advertisers.

Nashville Trip Country music stations can put a lot of net income on the books and get a lot of goodwill at the same time by sponsoring trips to Nashville. A travel agent or busline (within 500 miles of Nashville) will give you a package price. You advertise the trip as a special treat for your listeners. You can normally net 20 to 25% through discount or group prices/rates for transport, hotels, and joint sponsorships. 45 people on a $199 trip will net you about $2000 for a 60-day promotional effort.

Easter Hunt   The Easter hunt prize is a check for $100 in a plastic Easter egg hidden in a public place. Advertisers sponsor broadcast clues to the location of the egg. Each time a lucky egg is found, a new hunt is begun. One station in a California town of 4000 has five each Easter.

The longest running easter promotion in New York is the six-foot "Colieco Easter Bunny," a multiple location drawing. Bunnies cost about $41.95 (plus freight), including promotional materials. Some stations that have run this promotion for many years have injected some extra excitement by pinning an envelope to each bunny. Promotional announcements ask, "which bunny is the bunny with the money?" One winner gets an extra bonus of $100 cash.

Moonlight Madness   Most communities have moonlight madness sales. Does yours have an early bird sale? Merchants open for business three hours early, with storewide discounts each hour from 6:00 A.M. to 9:00 P.M. The discounts are the largest between 6:00 A.M. and 7:00 A.M., smaller from 7:00 A.M. to 8:00 A.M., and still smaller from 8:00 A.M. to 9:00 A.M. Restaurants tie in with early bird sale breakfast specials.

College Football Broadcast   This type of broadcast is probably available on several different radio stations in an area. Many times it is by the same sportscaster. A local station gets the edge by doing a 15-minute open line broadcast before the game. Listeners call in and guess the score. The winner is announced in the post-game broadcast.

Fire Prevention Week   This is observed the first full week in October. Insurance agencies and groups of insurance agencies sponsor public service announcements, done by the local fire chief. Information includes how wood-burning stoves, fireplace inserts, and space heaters have posed new fire hazards.

Beat the Freeze   Listeners guess the date of the first day on which the temperature dips to 32° Fahrenheit.

WLBC Trick or Treat House   Distribute $5 bills to ten houses in your area. The first youngster who asks, "is this the WLBC Trick or Treat House?" is given a $5 bill by the householder. Some stations sell the promotion. Youngsters are asked to say, "Is this the WLBC Pepsi-Cola Trick or Treat House?"

Trick or treat bags that glow in the dark with the station's call letters on them are excellent goodwill gestures and get call-letter exposure.

Window Night   Most downtown stores put Christmas decorations up on a specific day. From 6:00 P.M. to 9:00 P.M., listeners are invited to come downtown to see the Christmas windows and look at decorated stores. Specials can be included. A committee of nonmerchants judges the windows and awards plaques to the winning stores for their Christmas window decorating. Shoppers might be offered free refreshments at a convenient downtown location. Each store might offer a door prize.

The Day after Thanksgiving This is one of the two biggest pre-Christmas shopping days. Make it a special event by programming "Christmas music to shop by"—12 hours of Christmas music (9:00 A.M to 9:00 P.M.). One programmer scheduled his ads for the 15-breaks. He sold his three local banks the uninterrupted quarter-hours of music programming. The banks sponsored IDs at the beginning and end of each quarter-hour (about 15 seconds long) and recited various Christmas-shopper benefits of shopping at home. It was great PR for the banks with their retailer customers. Businesses pull their resources to produce a master "shipping guide" to which businesses in the community subscribe: announcing special sales, special shipping hours, etc.

Christmas Shopping Guide During the Christmas season almost everyone is a prospective shopper. The little businesses that can afford only a small schedule are apt to get buried by the big customers. Local businesses can pool their resources to produce a master shopping guide in which they announce sales, specials, and shopping hours.

Personality Breakfast

The objectives of a personality breakfast are as follows:

- to introduce a client's new food product to a target audience of commuters, using a radio personality
- to offer the station's listeners a special reward for listening and a memento featuring their favorite DJ
- to build store traffic and encourage consumers to sample a new product

How It Worked for WLS-AM and FM (Chicago) The station (WLS) worked with the regional association of Burger King franchisees to offer a Croissandwich® and coffee at a special price ($1.99). The coffee was served in a Larry Lujack (the DJ) mug at 181 Chicago stores. The promotion was aimed at Lujack's commuter audience of about 800,000. Burger King purchased time exclusively in the morning show to advertise its new morning product. WLS ran *liners* calling attention to the promotion around the clock.

The Investment Burger King bought a schedule of spots, produced, and paid for coffee mugs and point-of-purchase displays. WLS invested time for liners.

Results 45,000 mugs were distributed, thereby increasing store traffic and generating almost $90,000 in new sales. WLS had found another way to market Larry Lujack to his target audience and offered listeners a special reward for listening.

MAKING CONTEST PROMOTIONS WORK FOR YOU

There seems to be a giant wave of promotionalism sweeping America. The crest of this promotional wave now embraces an increasing number of state governments as participation in sanctioned lotteries abounds.

Promotionalism is effective because unique or unusual incentives will cause people to change pre-existing patterns of product sampling. *Sampling* extends from toothpaste to the evening news to the morning radio show. Properly executed, this activity increases audience, customers and revenues for the organization engaged in promotion.

Broadcast stations have embraced contest promotions as a way of increasing audience ratings. Contest promotions have all the elements of other promotions, i.e., advertiser promotion, station promotion and product/sales promotions. However, they have the distinction not only of adding present audience levels but increasing the time the target audience spends listening to the radio station. The objective might be to increase the audience of the morning show or a particular daypart.

Americans love to talk about winners, and the purpose of any contest promotion is to create talk. The question for broadcast promoters is what kind of promotion can my station become involved in that reflects this kind of talk and attention? Some of these questions are usually considered:

- Should the contest be *big* or *unique*? Unique in what sense?
- What kind of prizes should we include, cash, trips, houses, cars, shopping sprees, appliances, etc.?
- How do we qualify the contestants? Distribute numbered tickets through retail outlets? Call phone numbers at random? Have a drawing based on mailed-in responses? Mail out contest forms ourselves? Take the fifth caller? Give away prizes on the street or at a client's place of business, point-of-purchase?
- Can we have a client tie-in and have other media cover the contest?

How Do We Create the Talk and Attention?

The 20-second rule The contest elements must be simple. The incentive, ease of play must be explained within a 30-second produced promo, allowing 10 seconds for intro and outro; 20 seconds for the message. If the excitement and incentive cannot be explained in 20 seconds the contest promotion is too confusing.

Concept: Big or Unique?

A concept is big when a lot of money and time are allocated to disseminating the message, i.e., billboards, newspaper, direct mail, television or radio. Being big does not mean it is effective. Over the years many radio stations in heavily supported multimedia campaigns have given away large cash amounts and even houses that failed to produce any measurable increase in cume listening.

To create talk in the marketplace the promotional contest concept must be unique and provide vicarious enjoyment to listeners who are not regular broadcast contest players. The concept can be big and unique without tremendous collateral media support and create tremendous talk and substantial ratings increases as well. Two examples of this principle are "The Last Contest" and "The Great Race."

The Last Contest: This features hundreds of superbly produced and previously unimaginable prize packages that are described on the air (as spots) throughout the day for weeks. Each prize package is numbered and listeners are asked to remember the number of their favorite prize package and be listening for the mystery contest number. You will announce that number and ask for the 95th caller, for example, to claim the prize package.

The Great Race: This features the morning and afternoon disc jockeys engaging in friendly barbs that turn into a competition that is so intense that they ask the listeners to really one-up the other guy. The promotion culminates with a race around-the-world. Each disc jockey leaves town on the same day and time via commercial air carriers. One goes east and the other goes west. The first one back is the winner. Listeners are asked to guess who would win and what time and day they would return from their trip around the world.

The jocks would call in daily from the most exotic ports-of-call on the planet and report their progress and observations to the local audience. This program has, in the past, created lots of talk and the ratings of stations that did it soared in several markets.

When it comes to prizes, Jeff Lewis, President, Jeff Lewis Marketing, categorizes prizes based on perceptions of *wants* or *needs*. Cash is a prize that addresses listeners needs. Everyone wants more disposable income to pay bills or to buy something they feel they need.

Prizes that deal with wants are those items that most people would not buy for themselves if they had the money. Wants prizes are the kind of gift that people perceive as too frivolous or luxurious for them to use, have or enjoy, i.e., it is "beyond the reach of ordinary folks". Wants prizes include trips and cruises, furs, jewelry, luxury automobiles, or a shopping spree of designer label clothing—items they want to experience, but which they would be unable to consciously justify buying.

According to Mr. Lewis, contest promotions featuring large amounts of cash create strong initial impact but lack the sustaining quality of contests offering different wants prizes. Cash giveaways motivate the greatest number of contestants, but wants prizes create the most excited winners. The most consistently successful broadcast contest promotions have elements of both needs and wants prizes.

Your Target Audience

Contest promotions have a dual audience target—those who actively participate and those who participate by actively listening. The real objective of a contest promotion is to placate the actives with sufficient prize incentives and excite the passives as they enjoy the competition among the actives. It is not necessary to make winning easy but it is necessary that winning be perceived as simple, even to the point where a station drives contestants crazy and provides the audience with considerable entertainment.

Very often stations focus on the actives and neglect the passives. Promotions that "take the tenth caller" provide little excitement to the audience member who tends not to be one who calls. It thus serves as a turn-off and causes temporary boredom, anathema to ratings success. The greatest ratings impact comes from those contest promotions that cause talk and interest among the passives. The key to getting their attention depends on the on-air excitement of the produced promo. The kind of on-air promo that causes the passives to evoke vicarious pleasure responses like "How can they give away all those incredible prizes?" or "Gee, I'd love to win that." This audience group is more interested in actively perceiving that directly participating; and they create the most talk and ratings gains because there are more passives than actives.

NOTES

1. Emmerling, John. *It Only Takes One: How to Create the Right Idea*. Belmont, CA, 1989, p. 126.
2. 2B System Corporation. Madison, Michigan.

7
Ethics and Promotions Standards

ETHICS IN RADIO PROMOTION

Throughout history, business ethics have been ruled by societal ethics. Business owners and operators are frequently challenged by evolving ethical mores.

In the climatic scene of the hit movie, *Broadcast News*, William Hurt and Holly Hunter had the fight of their lives about, of all things, *ethics*. Hurt has faked a shot in a news report. Hunter finds it unacceptable.

"You could have been fired for that," she admonishes.

"I was promoted for it!" he boasts.

"You crossed the line," she reminds him.

"They keep moving that sucker [the line] . . . ," he justifies.

In four lines of dialogue we find the essence of a timeless debate about ethics. The American Heritage Dictionary of the English language (William Morris, editor, p. 450) defines ethics as the rules or standards governing the conduct of the members of a profession. In broadcasting, this has come to mean the principles by which we conduct ourselves as professionals, whether in broadcast journalism, broadcast programming, promotions, or life itself. Religion, society, and the professional world have come up with codes of laws, morals, and/or ethics to minimize the chaos that would exist were there no boundaries set for collective behavior. The code passed by the National Association of Broadcasters (NAB) has already changed. With or without changes, temporary or permanent, PDs and operators need guidelines.

Since radio promotion is not legally defined as advertising by government regulations, station owners, PDs and operators aren't overly concerned with the legal requirements for broadcast promotions. However, promotion departments always share ethical responsibility with other station personnel for the social effects of their programs and commercial messages. Violence and sexual exploitation are issues of broad social concern. Broadcast stations have enforced strict limitations (acceptance standards) on the content of commercials and promotions because carrying illegal material and running an illegal campaign could result in their not having a broadcast license renewed. Apart from this reality, broadcast stations need to to develop a positive image in the minds of their audience listeners—an image that encourages listening. Although most FCC and FTC regulations have an indirect bearing on promotions, several unethical practices by both television and radio stations persist.

Promotion people need the trust of their audience because they need the audience to respond to their messages. If your promotions bend, obscure, trick, or distort in order to entice listeners to tune in to your station, you will in the long run lose that audience. A short-term gain is not worth a long-term loss. Getting listeners to sample your station with flashy lures does a disservice to the program, the station, and the job of a PD.

But how do you pinpoint when your promotion has gone a step too far into the realm of deceit? "In most cases," according to Sy Cowles,[1] President of Cowles and Company Inc., "it is not easy. Mainly because we [promotions people] lack qualitative standards by which to judge our own work" (p. 169). The industry has no universally accepted definition of good promotion. In this case, good does not mean in the aesthetic sense, but in the way you measure the ability of your work to achieve specific goals.

That, Mr. Cowles points out, is the Achilles heel of promotion. We don't really know how to define a good campaign. We don't really know why some promotions work and why others don't. If a show gets great ratings, then the promo *may* have been good. If the ratings retreat, it *had to be* the promotions. Faced with this kind of uncertainty, there is a temptation to go for the most basic appeals in advertising and promotion: to find the broadest common denominator around which to build a campaign; to rely on elements that have the greatest known attraction to the greatest number of people: sex, violence, sloth, lust, greed, and even humor.

"But there is nothing wrong with that," Sy Cowles says. "Provided that the project you are promoting deals with *that* subject" (p. 171). That is, it is completely ethical for a promotion campaign to use sex as a primary audience bait, provided that the program being promoted deals with sex as the core element of the show. For instance, if Dr. Ruth is doing a program on sex and the senior citizen, Mr. Cowles believes a promotional announcement detailing what will be discussed in the program is ethical and therefore in good taste. Ethical promotion, however, is not necessarily good promotion. But the point is, you don't have to go beyond ethical boundaries in order to be good at your job.

WHEN IS A PROMOTION UNETHICAL?

A promotion is unethical when it fabricates a promotional premise for a program that doesn't deal with that concept, whether through sex or any other device. False and misleading information about contests is unethical. Deliver what you promise. A promotional campaign is unethical when it promotes the promoter rather than the product. When people talk more about your promotion than they do about the product you are selling, you've probably gone across *that line*. Ethics in promotion has a lot to do with striving for excellence. The process begins with a complete understanding of the product and its marketing needs. It ends with a campaign that relates the solution as closely as possible to the problem.

There are five practices (explained in the next section) relating to audience promotion that should be considered unethical despite their frequent violation: (1) tolerating *payola* or *plugola*; (2) using excessive *hypoing* to boost audience ratings artificially; (3) airing phony testimonials by celebrities, actors, or members of the general public; (4) the use of warnings about content unsuitable for children actually intended as come-ons; and (5) creating erroneous impressions of program content or conditions for winning prizes.

Payola and Plugola

Payola refers to illegal payment for promoting a recording or song on the air. DJs who accept free merchandise or other bribes for playing particular songs are guilty of taking payola. *Plugola* is a variant in which material is included in a program for the purpose of covertly promoting or advertising a product without disclosing that payment of some kind was made.

The dividing line between promotion and advertising becomes blurred when stations join with advertisers (in for example, promoting a rock concert). The penalties for violating payola or plugola regulations include fines of up to $10,000, a year in jail, or both for each offense. Fines have been imposed by the FCC in many cases. However, numerous cases of payola or plugola go undetected by the FCC.

Hypoing

Hypoing constitutes flagrant artificial inflation of listening in order to maintain audience allegiance and drive up ratings and advertising rates. During rating periods, stations enhance their schedules and increase promotion and advertising to boost ratings. On-air contests get extra attention, specials on exploitable subjects such as rape and drugs are common, and interviews are loaded with big name guests and tabloid subjects. (See Figure 12.)

These practices are widespread in the industry and seem to violate the spirit of pronouncements made by the FTC, the Electronic Media Ratings Council, and the rating services. Prohibited are on-air mentions of ratings, particularly with the station's call letters, and the conduct of field surveys during a rating period is heavily monitored. Scheduling documentaries or news interviews *about ratings* during this period is especially discouraged by the rating services.

Phony Testimonials

Using an actor to pose as a celebrity or public figure is obviously misleading, as is false testimony solicited by producers of spots (promos). Both practices may violate FTC regulations. Public display of microphones influences people's responses in subtle ways in "on-the-street" interviews, a popular means of obtaining contents for local promotional spots. Most people tend to give positive responses if they anticipate some reward such as a giveaway.

Misleading contests are to be avoided at all costs. Make certain that the contest rules and all advertising copies are absolutely clear and factually correct. The contest rules cannot in any way predetermine, prearrange, affect, or otherwise persuade the outcome of the contest. This is illegal.

Warnings

Advisories are intended to inform parents so they can restrict their children's listening. It appears, however, that many so-called warnings are in fact designed to draw adult audiences and have little effect on the numbers of children listening to the

Warnings to Avoid Hypoing

Survey Announcements

Relating to Survey Announcements/The National Association of Broadcasters (NAB) is ". . . concerned with the effects of the practice, engaged by some stations of exhorting the public to cooperate with radio ratings surveys" in progress.

The Electronic Media Rating Council (EMRC) opposes ". . . any attempt by stations to exhort the public to cooperate with radio audience measurement services whether over the air or by any other means, and recommends to syndicated audience measurement services that the practice be discouraged because of its possible biasing effects." The EMRC has amended its minimum standards to define Survey Announcements as biasing.

The American Association of Advertising Agencies (AAAA) ". . . opposes any attempt in any medium to exhort the public to cooperate with any audience measurement survey by calling attention to such research by any means."

The Arbitron Radio Advisory Council has reiterated its ". . . continuing opposition to rating bias in the form of on-air survey announcements" and "off-air announcements that call attention to the scheduled survey dates or diary methodology. This opposition includes direct mail, newspapers, or . . . advertising."

FTC Guidelines

The FTC Guidelines Regarding Deceptive Claims of Broadcast Audience Coverage point out that radio stations ". . . should not engage in activities calculated to distort or inflate such data—for example, by conducting a special contest, or instituting unusual advertising or other promotional efforts designed to increase audiences only during the survey period."

It is the opinion of Arbitron that while many stations may engage in promotional activities during a survey period and may not be attempting to distort audiences, some stations may conduct their promotional activity for the specific purpose of biasing or distorting audience estimates during the rating period. Such activities could affect the behavior of survey participants, thereby changing audience estimates from what they would have been if no such activity had been conducted during the survey period.

Arbitron Radio urges all broadcasters not to engage in any of the above activities.

▶ *Figure 12* Hypoing constitutes a violation of FTC guidlines. Stations are issued these guidelines which are unenforceable. Courtesy Arbitron.

program. These warnings do not change the content of the program as they are targeted for specific audiences. This is why children are warned to keep away.

False Impression

Creating an erroneous impression in on-air promotion is unethical. Continued discrepancy between what is promised in on-air promos and what is actually delivered eventually causes the audience to tune out. A more serious concern is misleading information about on-air contests. The FCC requires broadcasters to state periodically all the terms of a contest on the air. Continued misleading impressions can lead to investigations by the FCC and are prohibited in federal regulations.

The subject of ethics for promotion usually involves questions of personal and social values rather than purely legal issues. Section 73.1216 of the FCC's[2] rules and regulations outlines the do's and dont's of contest promotions.

- Stations are prohibited from running a contest in which contestants are required to pay in order to play.
- The FCC regards as lottery any contest in which the elements of prize, chance, and consideration exist. (See Chapter 4, for explanation.) In other words, contestants must not have to risk something in order to win.
- The FCC requires any licensee that broadcasts or advertises information about a contest that it conducts to *fully* and *accurately* disclose the material terms of the contest. You must conduct the contest as advertised.

What are the material terms of a contest? This means you must disclose the following information about each contest:

- how to enter and participate in the contest (e.g., "Go to Brian's Bake Shop and pick up an entry form.")
- eligibility requirements and/or restrictions (e.g., "Contest is open to all listeners 18 years of age and older. Employees of WLBC, Brian's Bake Shop, and the immediate families are ineligible.")
- entry deadlines (e.g., "All entries must be postmarked by midnight, January 15, 1992.")
- whether prizes are to be won and when those prizes will be awarded (e.g., "There will be a grand prize, a second prize, and a third prize awarded. The three winning entries will be drawn on July 23, 1992.")
- the time, place, and means of selecting winners (e.g., "Winners will be selected by a random drawing at midnight July 23, 1992, by . . .")
- the extent, nature, and value of the prizes (e.g., "The grand prize is round-trip air transportation to Sierra Leone. Total value of prize is $723.49.")
- tie breaking procedures (e.g., "In the event of a tie, the judges decision will be final.") (See Figure 13.)

Alan Batten,[3] Director of Advertising and Promotion, WSOC-TV, Charlotte, North Carolina, has made the following checklist for planning a contest:

1. Does your copy mislead, overstate, or over-hype the exact nature or value of the prizes?

2. Do the rules provide safeguards to assure fair opportunities for all contestants to win the announced prizes?

3. Do you tell the public the exact times and places for the drawing and awarding of the prize, and is this within reason the shortest period of time after the close of the contest?

4. Does the copy imply that listening to your station and going to the sponsor's place of business will give the contestant a better chance to win?

5. Does the copy accurately relate the rules and conditions of the contest and do you plan to repeat this all through the contest?

6. Do you plan to disclose any changes in the contest rules to your listeners (the public).

CONTEST RULES and REGULATIONS

1) YOU HAVE TWO WAYS TO ENTER TO WIN CASH AND/OR PRIZES FROM WHTT. DISPLAY THE LARGE PART OF THIS BUMPER STICKER ON YOUR VEHICLE AND FILL OUT THE OFFICIAL WHTT ENTRY FORM AND MAIL IT TO: WHTT-HITRADIO, BOSTON, MA 02199. ENTRIES MUST BE MAILED SEPARATELY AND RECEIVED BY DECEMBER 24, 1984.

2) IF YOUR VEHICLE IS SEEN DISPLAYING THE WHTT BUMPER STICKER IN THE WHTT GREATER BOSTON LISTENING AREA BY OUR SPOTTERS, THEY WILL WRITE DOWN YOUR LICENSE NUMBER AND THE MAKE AND MODEL OF YOUR VEHICLE AND ENTER THAT INFORMATION INTO THE PRIZE DRAWING.

3) SELECTED DAYS NOW THROUGH DECEMBER 31, 1984, WHTT WILL CHOOSE LICENSE NUMBERS BY RANDOM DRAWING FROM ALL THE OFFICIAL ENTRIES RECEIVED, AND, FROM THE SPOTTER'S INFORMATION COLLECTED, AND ANNOUNCE THE SELECTED LICENSE NUMBERS ON THE AIR. WHEN YOU HEAR YOUR LICENSE NUMBER, YOU HAVE 103 MINUTES TO CALL US AND IDENTIFY YOURSELF AND YOUR VEHICLE. YOU THEN MUST BRING YOUR VEHICLE REGISTRATION TO THE WHTT OFFICE FOR VERIFICATION TO CLAIM YOUR PRIZE. PRIZES NOT CLAIMED WITHIN 30 DAYS WILL NOT BE AWARDED.

4) ONLY ONE PRIZE PER LICENSE NUMBER, PER MONTH WILL BE AWARDED. CBS EMPLOYEES AND THEIR FAMILIES ARE NOT ELIGIBLE TO ENTER. ONLY REGISTERED OWNERS OF PRIVATE, NON-COMMERCIAL VEHICLES ARE ELIGIBLE. THE DECISION OF THE JUDGES IS FINAL. VOID WHERE PROHIBITED.

5) THIS IS A FREE BUMPER STICKER. NO PURCHASE REQUIRED TO ENTER. FREE BUMPER STICKER AND ENTRY FORM AVAILABLE WHILE SUPPLIES LAST AT THE WHTT OFFICES, OR BY SENDING A SELF-ADDRESSED, STAMPED, BUSINESS SIZED ENVELOPE TO WHTT BUMPER STICKER, 4418 PRUDENTIAL TOWER, BOSTON, MA 02199. WRITTEN REQUESTS MUST BE RECEIVED BY DECEMBER 1, 1984.

OFFICIAL W-H-T-T ENTRY FORM
(Please print)

NAME _____ AGE _____
ADDRESS _____
CITY _____ STATE _____ ZIP _____
HOME PHONE _____ WORK PHONE _____
CAR MAKE & MODEL _____
LICENSE NUMBER _____ STATE _____
HOURS OF DAY YOU LISTEN TO WHTT _____

mail this entry to: 103FM WHTT HITRADIO
BOSTON, MA 02199

▶ *Figure 13* Stations are obliged to make contest rules clear to the public. Courtesy WHTT-FM.

7. Are all entries consistently judged by the same standards?
8. Do your rules and procedures provide for adequate hands-on supervision by the station staff?
9. Does your copy imply, cloud, or convey false clues or information about the contest?
10. Are you sure that the contest does not provide for secret assistance or "super clues" obtainable only by a select few.

The public must not be misled concerning the nature of prizes. Specifics must be stated. Any changes in contest rules must be promptly conveyed to the public.

Rigging contests—for example, predetermining winners in advance—is a serious violation of the law and can result in a substantial penalty that could include license revocation.

It is a good practice to maintain all pertinent contest information, including signed prize receipts and releases by winners. This could prevent grave problems and protect your station should a conflict arise later. Stations that award prizes valued at $600 or more are expected by law to file a 1099 Misc form with the Internal Revenue Services.

Stations incur no tax liability by doing this, but failure to comply can put your station at odds with the law.

PARTING WORDS ABOUT CONTESTS

How do you develop that great idea and not end up in jail? Bingo! Contests! What was at one time a constant fixture of radio stations is today a constant diet of television stations and cable systems. After you have determined that a particular contest is going to increase your ratings three or four points, there are some very important steps to take. Steps that could save your job and your station's license.

Check with Your Lawyer

In today's business climate, it is almost impossible for any station to operate without some form of legal counsel. Since each state has slightly different laws, the general guidelines in this chapter may not apply to your ADI (area of dominant influence). Be sure your GM and lawyer see and approve each contest idea before you broadcast it or put it in print.

Lottery or Contest?

In addition to being a ratings booster, a contest is defined as any plan that offers a prize to the public based on chance, knowledge, or skill. While your contest might contain one or two of these elements, if it has a prize, (like most contests) chance, and consideration, it is an illegal lottery.

In the United States, there are legal state-run lotteries that have been exempted by the FCC from the illegal lottery rule. Anything else is an illegal lottery and broadcast of one would be in violation of your station's license.

A lottery is defined as any scheme for the distribution of prizes by lot or chance by which someone, paying money or giving anything of value to another, obtains a token that entitles him or her to receive a larger or smaller value as some formula of chance may determine. The definition of lottery may vary from state to state, country to country. If in doubt, contact a lawyer.

One element that has been quite ambiguous in its definition is *consideration*. Consideration ranges from making payments, buying tickets, paying to enter an event where entry forms are available, requiring participants to purchase items or requiring participants to spend considerable time and money. If a contestant must give you anything of value, including time, in order to participate, then consideration may be present.

Clearances

Now, you have run a successful contest. You seem to have followed all the rules. But don't forget! If the contestant's picture, voice, or other identifying materials are to be used in connection with the promotion of the contest, first obtain a full model and rights release. When in doubt, consult your lawyer.

NOTES

1. Cowles, Sy. "Network Television Promotion" in *Promotion & Marketing for Broadcast and Cable*. Prospect Heights, Illinois, p. 169.
2. Federal Communications Commission. *Code of Federal Regulations*, Section 73.1216. Washington, D.C.: Government Printing Office.
3. Batten, Alan. "Contests: Some Dos and Don'ts" *BPME Image*, January 1986, p. 26.

8
Radio Promotions and the Future

RADIO: THE GROWTH MEDIUM

Critics argue that the risk-takers and innovators of radio are being replaced by number crunchers and technocrats. The industry *is not* losing its adventurers and pioneers. Although innovative forms are changing, the spirit in radio remains the same. Today our best and brightest owners and managers are designing new forms of financing and ownership. They are challenging the industry with new standards for station performance, and they are measuring their own successes by different criteria. In terms of programming, technology, and profitability, there is a pool of young dynamic talent in radio that is building upon the inherited legacy of radio's founding generation.

The generational change in radio coincides with another major change in the way we do business: changes in economics and government regulation are effectively making a two-tiered industry. On one level, we have the radio station operators who plan to own their stations for long periods of time and who see themselves as traditional broadcasters. On the other level, station traders see radio as an arena for buying and selling. For traders, operations are important only in so far as they sustain or increase a station's purchase price or profit potential. Traders tend to be major companies with deeper financial resources, who will seek to improve market position through trades, and who look to radio for strong returns on investment over 5- to 10-year periods. While operators and traders are mutually exclusive, they work with two different sets of assumptions.

Consider this: In all of 1984, 438 radio stations were sold in this country. During the first six months of 1988, 542 radio stations changed hands. Several recent regulatory changes have had a profound impact on the market for radio stations. The Federal Communications Commission (FCC) advanced tax incentives to promote minority ownership, eliminated the 7-7-7 Rule (raising the limit on the number of stations one person may own), and removed the prohibition against trafficking, [buying stations to make a quick sale and quick profit without any intent of keeping the station for it's programming] which required stations to be held at least three years before sale. (See Figure 14.)

Historically, station purchase prices comprised both present and future values. The decision to buy is made on the basis of the facility and the market potential. As a result, record high prices are changing a station's debt-to-equity ratio (Figure 15).

Radio Promotions and the Future 75

More Than 9,200 Commercial Radio Stations Are On The Air Today

Year-End	AM Stations On Air	FM Stations On Air	Total Stations On Air
1989	4,975	4,269	9,244
1988	4,932	4,155	9,087
1987	4,902	4,041	8,943
1985	4,718	3,875	8,593
1980	4,589	3,282	7,871
1970	4,323	2,196	6,519
1960	3,539	815	4,354

▶ *Figure 14* *The number of radio stations in operation shows a dramatic growth from 1960 to 1970 and a steady increase from 1970 to 1989. Courtesy Radio Advertising Bureau.*

Nonbroadcasters investing in radio are attracted to resale profits at the expense of operations. This has been the pattern. As stations change hands, each new owner pays higher prices and assumes an even greater debt-to-equity ratio. This trend has overvalued radio properties and is preventing would-be financiers from entering the acquisition marketplace.

The first and most immediate challenge for radio operators is to revitalize, reposition, and remarket the AM band for listeners and clients. This could be accomplished by introducing continuous-band tuners, distributing AM stereo radio receivers, and using compact discs and digital audio fidelity on both broadcast bands. Some programmers are experimenting with new formats that include comedy and game shows.

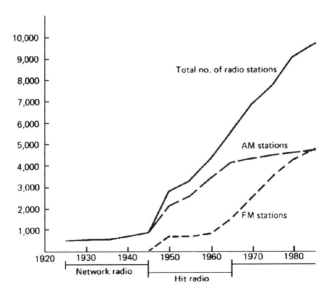

▶ *Figure 15* *The number of on-air radio stations, 1920 to 1985. Courtesy Broadcasting Yearbook, Broadcasting Marketing Technology.*

The second challenge is for radio operators to become more innovative in program production and distribution, while achieving new levels of operational efficiency and profitability. This must be accomplished without sacrificing localism or the ability to program directly to the needs of specific market segments.

Third, radio operators must take advantage of developing technology. This will further develop the consumer's use of media. It also will present the radio industry with opportunities for distribution on cable or the creation of even narrower radio program formats. This might mean having radio marketers link their services to computers and satellites as a means of reaching specific customers.

Fourth, radio must adapt to the aging population. This means developing new programming formats. The fact that the Beatles are "middle of the road" music for today's teenagers shows that new ways of programming and using radio are inevitable. In addition, the aging post-World War II baby boom generation and two-income professional couples will be spending discretionary income and leisure time in different ways. Radio will change with them.

Radio will sound different as more diverse groups gain access to broadcast it, and as programming targets become smaller and even more specialized. There will also be changes in the number of commercials sold and the way they are sold.

On the local level, car dealers, banks, fast food restaurants, soda bottlers, beer distributors, and retail stores will continue to be radio's most important clients. Radio's ability to attract more revenue from local businesses has been and will continue to be a direct function of the economy and the advertisers ability to bring in a profit. Tight margins, excess inventory, pending and yet-to-be thought of changes in tax codes and the level of consumerism, compared with growth of disposable income are the variables that will effect radio's ability to sell commercial time. (See Figure 16.)

TRENDS

Promoting Radio on Television

Change is an integral part of the American experience, yet usually it is only reluctantly accepted by marketers. Today's broadcaster's scope is no longer limited to competing for traditional audiences, with traditional competitors using traditional marketing strategies. Every opportunity should be evaluated on its own merit, not by tradition (or what has been). All marketing plans must contain a strategy for using other media. Newspapers are playing a decreasing role in broadcast promotion.

Although many radio stations consider it too expensive, television advertising increasingly is a major medium for reaching target audiences in increasing numbers. Because of the cost involved, syndicated campaigns are available to reduce costs. Most media will consider trading time between radio, television, and cable, either on a straight promotion basis or part trade–part cash. Most radio formats can be sold very effectively on television in 30 seconds, and the medium is excellent for creating awareness and providing a sample of new programs to large numbers of viewers. Many station are now using animation, graphics, and laser art to stress the look and feel of their sound. Because of the expense, more radio stations will do spot campaigns, which have created specific radio formats in syndication.

Year		Net	Spot	Local	Total
1965	$	60	275	582	917
	%*	7	30	63	
1970	$	56	371	881	1,308
	%	4	28	67	
1975	$	83	436	1461	1,980
	%	4	22	74	
1980	$	183	779	2740	3,702
	%	5	21	74	
1984	$	305	1135	4420	5,905
	%	5	20	75	
1990	$	478	1589	7998	10,065
	%	5	16	79	

▶ *Figure 16* Radio revenue trends show a major concentration of spending at the local level. Courtesy Radio Advertising Bureau.

According to the Television Advertising Bureau (TVB), radio spent approximately $100 million in 1988 on television advertising (up 9% from 1987). In 1989, according to the RAB, radio stations spent $78.9 million on spot TV and $154,000 on cable TV. These figures do not represent all of the advertising time traded by radio and television stations. The Pollack Group[1] in Pacific Palisades, California, believes this money was "poorly spent" (p. 18).[1] They, therefore, have developed some concepts that should be important to those PDs planning to use television to promote radio:

1. Stick to one point. Viewers of a television spot may not tune to your station until the following day—a time span in which your message can be lost or forgotten. Viewers should come away with a clear point that stays with them.

2. Be creative. There are other competing images and commercials before and after your spot. Only the most creative and best in terms of execution and impact will break through this maze and become implanted in the viewer's mind.

3. Use music beds. Your spot should have a music bed that matches the kind of music listeners will hear when they tune in to your station—unless, of course, you are a News/Talk station.

4. Don't overdo it. A spot that is too entertaining doesn't sell the product. Develop one clear message, and then reinforce it with your on-air imaging and promotional activity.

5. Hire the right people. Work with a creative agency or production house that understands your station and the message you want to get across.

6. Allocate funds wisely. Don't design a TV campaign that costs $50,000 and then only have $20,000 to spend for placement. You are wasting your money. Spend your money wisely. The same commitment you have to choosing the right records for your station also must be extended to promotion and marketing plans in order to ensure market leadership.

KEYS TO TARGET MARKETING FOR THE 1990S

Not only do Americans have too many obligations, there are too many options. The average American is hit with 5000 advertisements a day, 24-hour shopping is as close as a telephone, 200 or more new products line supermarket shelves every week, television delivers at least 15 channels to most households, and about one third of all Americans now watch two or more programs at a time.[2]

Promotion in the 1990s has changed dramatically as have the media. Two convergent trends have changed things for all of us: time and technology. Time is one of the most precious commodities. Nobody has enough time. 74% of those aged 35–44 feel time pressure. In 1967, at a Senate subcommittee testimony, it was predicted that by 1985 the average person would work 22 hours a week or 27 weeks a year and would be able to retire by age 38. This is not so. Leisure time has *shrunk* by 37% since 1973. The work week, including commuting, has increased from 41 hours a week to 47 hours a week.

Time has become precious and the search for it is vicious. We now have all these labor-saving devices—dishwashers, washing machines, clothes dryers, and frost-free or self-defrosting refrigerators—that are designed to save time. But all they have done is raise our standards for ourselves. In this decade, the time pressure and constraints will get worse.

The other element is new technology. In radio today we are talking about digital technology. Digital is to analog what the transistor is to the vacuum tube. Digital has changed everything for the future. What happens with a digital stream is that we can put a lot more information in the same spectrum space. 1986 ushered in CDs (compact discs), HDTV (high definition television), digital audiotape (DAT) computers attached to television sets, and cable audio.

With digital products, we were introduced to a system that is able to send multiple signals in the same space in less time. The multiscreen television sets are here. With computerized television, we are able to roam and search for the programming that we want, when we want it. We don't have to go through all 88 channels that appear on our cable systems. Listeners and viewers can now roam, pursue, and capture the programming they want. This makes the PD's job a little more difficult. Content is getting to be the determinant of what gets listened to. In the future, *real* time will not exist in broadcasting; most programs will be listened to or viewed in *delayed* time. Programs will be recorded by listeners and played back when the listener wants it, at his or her own time. Because of this, personal choices are becoming much more important. Now there is emphasis on identifying lifestyle groups, consumer groups, and also identifying their behavior as a result. For the broadcast programmer this means more and more targeting and more and more special promotions.

To survive, radio programmers and PDs have to reach out, select, and involve people. Stations must do a little research—find out who their listeners are, what they want to listen to, and deliver such programming to them. More importantly, once the programming has been delivered, the PD should go all out to *tell* listeners about it. With 88 TV channels, countless radio stations, and digital radio that comes in via satellite directly to the home (DBS), the audience is not going to be able to find your station on their own.

One-to-One Marketing

Three areas are being widely discussed as a means of breaking through this maze to reach target listeners:

Direct Mail Direct mail refers to mass mailings, usually by commercial companies that specialize in direct mail. This is an incredible growth medium, which has used items such as activator cards, listener cards, focussed mail, and program guides. By using services such as Arbitron's "Fingerprint", a joint product of Arbitron and radio programming consultant Gary Donahue that tells what competing stations with similar formats are doing in each demographic area, radio stations can target specific demographic groups (usually by zip codes) with direct mail campaigns. A station's advertiser may sponsor a secret contest that only direct mail recipients know about. The mail may say that the first person to call the station after a specific song or during a specific time period will receive a cash award—although no mention of the prize is made on the air.

The cable industry has used this medium very effectively. Direct mail or one-to one marketing, according to Kathren John,[3] of Direct Marketing Results, achieves the highest percentage of response per thousand people reached.

Why Does Direct Mail Work?

Direct mail is a tangible advertising medium that allows the time and space to properly position your station. You can reach out and ask a potential audience to "sample" your station. Direct mail works when targeted to the proper demographic group. It would be unwise to send a mailing about your hard rock station to a retirement community. Direct mail increases your station's exposure and broadens your listener base. It can be a two-way communications effort by not only telling your story but, on request, listeners can respond to a number of questions about themselves and your station. An activator card might ask the following questions:

1. Do you listen at work?
2. What hours of the day do you listen?
3. Do you listen during your morning drive?
4. Where do you listen?

With this information you can go in and focus on your morning drive or midday listeners and bring them into other aspects of your programming. Direct mail can include coupons that will lower production costs. You can recruit sponsors and include their coupons in your piece of mail. A sweepstakes can bring in new customers to your clients. If, for example, a direct mail piece costs $75,000 to do, you might be able to liquidate $30,000 of it by getting your clients involved.

WTPI (Indianapolis), a few years ago, did a direct mail/telemarketing campaign during which they contacted various members of their potential listening audience. One potential listener called up the station to say she never realized that her radio dial went beyond 103.5 (WTPI is at 107.9). There are people who do not know that your station exists. Direct mail can stimulate listeners to sample your station if you give a reason to listen.

An advantage of direct mail is that it places you right in front of your audience without competition or distraction.

With the amount of mail people receive each day, how do you get a potential listener to *open* your mail?

- Make the offer attractive.
- Get their attention immediately and hold it.
- Use bold attractive lettering.
- Address the letter vertically. It's unusual and draws attention.
- Ask a provocative question. "Are you daring enough to try WLBC?"
- Make an astounding statement. "Go ahead open it, it's different".

It also helps to maintain contact with your core audience members to let them know what your station is doing, new promotions that are coming up, new services and programs that are planned, etc. Remember while doing this that there are four or five other stations in your market involved in aggressive television, telemarketing, and direct mail campaigns to attack and win over your core audience members. There is nothing like a *loyal* audience. They will listen to you as long as you continue to provide the services they want.

Telemarketing This refers to telephone sales and is often directed at in-office listeners. These "in-office" listeners have become very important to radio stations, especially during the midday daypart, and raising midday listening can greatly influence afternoon drive ratings. This kind of listening is also known as *forced listening*, listening over which the listener has no control. This occurs in supermarkets, shopping malls, car pools, workplaces, and retail outlets. Telemarketers often reward those offices and retail managers who control radio station selection at their place of work or business. No mention of these rewards is made on the air if the offer is connected with a *secret contest*. PDs may randomly call businesses listed in the yellow pages and offer cash reward and on-air mention (plugs) to the manager, such as "Thanks to Folarin's Auto Repair for listening to WLBC 96." Someone from Folarin's would then have a limited period of time to call in and claim the prize. This kind of telephone selling guarantees that your message reaches your target audience by direct telephone conversation with its members.

With telemarketing you have a chance to adapt or readjust your message by the reactions of respondents. You are able to react personally (immediately) with the listener. You can discuss such things as the listener's likes and dislikes, and you are able to correct misconceptions about your station.

Telemarketing creates two-way communication and allows input from your audience or potential audience. This gives you an opportunity to answer questions and hear objections and observations about your station.

Data Base Management

This helps the station bond with its audience. It creates a feeling (among your audience members) of belonging to a club or family. The days are gone when there were four or five radio stations in a market, when *Look*, *Life*, and the *Saturday Evening Post* were the three major magazines, and there were three channels or networks. Data base marketing can help create an affinity with your listeners. Data base management allows you to target your marketing much more closely and effectively than a mass mailing. With data base you can begin to develop a relationship between the listener and the station. An *activator* card, properly answered by your listener, can help you identify your core audience. It can tell you *who* listens, *when* they listen, and *what* they like or dislike about your station. It can also help you find our about the kind of people who listen to you: their occupations, education, income levels, pastimes, etc. You could send birthday cards on their birthdays, or just have the day mentioned on the air. It could help keep your core audience.

How Can Your Core Audience Increase?

Let's assume your station has a core audience of 6000. Through research, learn all you can about the general characteristics of these listeners. Who are they demographically and psychographically? What are they all about? Why do they listen to your station? Keep a computerized record of this information. Cross that profile against your entire area and find additional people who are like your core audience and start mailing to them. If they correctly mirror your core listeners, they also are potential listeners for your station.

RADIO AND THE NEW TECHNOLOGY

Digital audio is dramatically changing radio broadcasting technology. The benefits of digital audio are immediately apparent to both radio broadcasters and the general public. This is evidenced by the (almost instant) acceptance of CD. The higher sound quality displayed by CDs is the principal reason that radio stations have now begun to seek better ways of delivering a higher quality product to their listeners.

There are basically three generations of digital audio systems that are currently in various stages of integration into the radio station. The first generation is the CD and digital audiotape (DAT); the second generation is the digital audio workstation (DAW); and the third generation will be the complete integration of all digital computerized systems to work towards the radio station of the future. Of prime interest to radio programmers and PDs, at this time, is the DAW.

The Digital Audio Workstation

Until recently, most radio stations' production, in-house promotion, and commercial production were done in the analog mode using audiotape. This is rapidly changing, however, with the introduction of the DAW which has a much larger scope than either DATs or CDs.

The DAW consists of an entire computer system with video screen; recordable, high-density Winchester hard drives; vast amounts of random access memory; customized software; and recordable optical discs. The optical disc, according to Paul Donahue, Director of Engineering, Gannet Radio, is similar to the CD, however the CD is a 5-inch disc that is currently used for playback only. The DAW however, incorporates an 11-inch optical disc, which has four times the storage of a CD and has the capability to record *and* edit digital audio.

The DAW brings a simpler operation to in-house production, promotion, and commercial production, while using the higher quality of DAT in the recording and editing mode.

With this system, radio stations are able to record and edit music, changing it in ways that allow the radio station to sound more creative. This is how it works. DAW uses a video screen and a hand-held mouse, much like a Macintosh computer, to edit the audio. Gone are the days of using a tape, a razor blade, and a splicing block to cut and paste various pieces of a commercial or promotion on analog tape.

Because the DAW operates like a word processor, an operator can see on the video screen a picture of the audio to be edited. Then, using the hand-held mouse, the operator can move and edit the sounds much like operating a reel-to-reel tape recorder. This allows the station to deliver a better quality product to its listeners and advertisers. The final advantage of the DAW is its productivity gains, which initially appeared to be between 25 and 30%, by using the on-screen cut-and-paste method. This initial 25–30% gain in reduction of time spent on producing commercials and promotional materials is a major factor when considering the cost of talented production people.

The Phone-Patch Voice-Over Concept

Another development in radio's technological future is the introduction of fax, or telephone interconnect, and overnight delivery. These allow stations everywhere to have promos voiced in Los Angeles with state-of-the-art equipment by top talent using only the phone on their desk.

How It Works The client calls to set up a time when the talent is available. The talent voices the promo; the client directs. The client faxes a script, purchase order, and other information. The talent prepares the script for the session. The producer calls the talent (as per arrangement at a mutually convenient time). The talent is ready and stands at a mike with headphones on. With the telephone interconnect, the talent turns on a switch that answers the phone. Through the right ear of the headphones, the talent hears the producer and the actual recording as it feeds into a tape recorder (there is no engineer involved). The producer can also hear everything that's going on because of the high quality of the phone-patch.

Phone-patch voice-over recording makes it possible for producers all over the country to work quickly and easily with voice-over talent located elsewhere. "This is the true future of voice-over. It's replacing a trip to the studio, the way the fax has replaced a trip to the post office."[4] says Paul Trembley, a voice-over actor, who has voiced promotions for CBS-TV and others.

PROMOTION DIRECTORS AND DEPARTMENTS OF THE FUTURE

The single most important change in promotion departments will be in the increased skills and professionalism of the PD. In addition to the basic skills already discussed in previous chapters, the PD of the future will need to have a greater understanding of and experience with more sophisticated marketing techniques.

As the marketing of radio stations increasingly employs tactics involving the combined use of direct mail, telemarketing, interactive computer and phone systems, and research (particularly the Arbitron "Fingerprint"), potential PDs will have to learn to integrate the use of these tools into their working vocabulary. Winning stations will hire PDs with the ability to contribute ideas at the strategic level and with the skills to execute at the tactical level.

The office of the future will be making even greater use of computers, and radio will be no exception. Tasks such as word processing, data base management, particularly for frequent-listener programs and project management, can all be handled faster and more accurately with today's computer. The PD of the future will be constantly looking for new techniques and tactics that will provide a competitive edge, one that will be an integral part of the station's SMP process.

Marketing radio stations today requires a broader perspective and a greater range of knowledge and skills than ever before. As more stations appear on the spectrum, radio marketers have a greater burden to differentiate, distinguish, and describe their audiences in order to sell the medium.

PDs are uniquely positioned to translate the programming objectives of a radio station into basic selling tools. Graphically, PDs put together valuable sales tools and design station coverage maps. The more they understand about the sales process, about the conditions in the marketplace, and about their clients, the better prepared they will be to make meaningful contributions to the marketing process.

Without strong basic marketing skills, the designation of PD will not survive as a position at radio stations by the end of this decade. With the acquisition of all the needed marketing and sales skills, some are predicting that the title of PD will be changed to marketing director.

NOTES

1. The Pollack Group. *BPME Image*, August/September, 1989, p. 18.
2. NAB Convention Notes. Las Vegas, Nevada, March, 1989.
3. John, Kathren. *BPME Image*, October, 1988, p. 36.
4. Trembley, Paul. *BPME Image*, October, 1988, p. 19.

Glossary

ABC American Broadcasting Company network. Was formerly NBC blue. Became ABC in 1945.
Account executive Station or agency salesperson.
Actives Listeners who call radio stations to make requests and comments or in response to contests and promotions.
AFTRA American Federation of Television and Radio Artists; union comprised of broadcast performers, announcers, DJs, and newscasters.
Aircheck Tape of live broadcast
AM Amplitude Modulation; method of signal transmission using Standard Broadcast band with frequencies between 535 and 1605 (1705) kHz.
Announcement Commercial (spot) or public service message of varying length.
AOR Album-Oriented Rock radio format.
Arbitron Audience measurement service employing a seven-day diary to determine the number of listeners tuned to area stations.
Audio A term used to denote the electrical signal that transmits sound.
Audio promo A voiced tag for an upcoming program usually placed over closing credits of preceding program.

Barter Exchange of airtime for programming or goods.
BEA Broadcast Education Association. Represents hundreds of colleges and universities with broadcast curriculums.
Beautiful Music Radio format comprised of low-key, mellow, popular music, generally with extensive orchestration.
Bed Music behind voice in commercial.
Benchmark Research conducted before the start of a promotional campaign to determine what the audience perceives at the time; provides a comparison for research conducted after a campaign to isolate the probable effects of the campaign.
Billboard Highway display sign or brief on-air listing of participating advertisers at the open and close of a program.
Birch A rating company that surveys radio stations in large and medium markets. It provides clients with both quantitative and qualitative data on local listening patterns, audience size and demographics.
BPME Broadcast Promotion and Marketing Executives. Major trade organization of broadcast and cable promotion and marketing directors.

Glossary

Bumper On-air tease preceding a series of commercial spots that encourages viewers or listeners to stay tuned to upcoming programming (uses such copy as "stay tuned for . . . right after these messages.")

Call letters Assigned station identification that begins with "W" for stations east of the Mississippi River and "K" for stations west of it. (Except KDKA, Philadelphia. The first radio station).

Call-ins People who telephone the station in response to on-air calls for listener response.

CBS Columbia Broadcasting System. Network formed in 1927. Originally, UBI (United Independent Broadcasters). Purchased by William S. Paley in in 1928.

Commercial Paid advertising announcement. Also known as *spot*.

Compact disc A disc that looks like a small silver phonograph record; it is read by laser beam and offers high-quality sound, with some signal degradation due to wear.

Copy Advertising message, continuity, script.

Copyright Registration of television or radio programs, movies, or other media with the Federal Copyright Office, restricting permission for use.

Copywriter One who writes commercial or promotional copy.

Crossover Use of a character or personality from one program in another.

Dayparts The three periods or segments of a broadcast day divided as follows: 6:00–10:00 A.M.; 10:00 A.M.–3:00 P.M.; 3:00–7:00 P.M.

DBS Direct broadcast satellites—programming going from satellite to home receiving dishes, bypassing a ground-based distributor, such as a broadcast station.

Demographics Audience statistical data pertaining to age, sex, race, income, etc.

Disc jockey (DJ) Host of radio music program or an announcer.

Downscale Audience with lower than average socioeconomic demographics, especially low income.

Drive time Radio's version of prime time, basically, those dayparts in which many listeners are in automobiles—early morning and late afternoon.

Easy listening A music format featuring mellow songs and instrumentals.

FCC Federal Communications Commission: An arm of the United States government responsible for overseeing of broadcast communications and certain other types of communications.

Fingerprint A programming guide that tells how well a market's competing stations with similar formats are doing in each geographic area as identified by zip code.

Focus group Group of a dozen or so people assembled by researchers to elicit reactions to a program, personality, promotion, or commercial advertisement.

Format Type of programming a station offers; arrangement and selection of a certain type of music, tone, announcing approach, and other materials to reach a specific target audience. The formula.

Guides   Program listings, published in printed or electronic form.

Headphones   Speakers mounted on ears, headsets, cans.
Hype   Excessive promotion.
Hypoing   Boosting audience size during a ratings period through greater than usual amounts of promotion.

ID   Hourly identification of the station's name, network, or system; may include frequency, may be a jingle.
Identity   In promotion, the actual program content and community service of a station, network, or system, which may or may not be congruent with its public image.
Image position   Promotional campaigns intended to foster a positive attitude in the public's mind toward a particular station, network, or system.
Independent   A commercial radio or television station that is not affiliated with a broadcast network.

Jingle   Musical commercial or promo that is a station's signature and includes call letters and frequency number.

Liner cards   Written on-air promos used to ensure adherence to station image; prepared ad-libs.
Live copy   Material read over the air, not prerecorded.
Live tag   Postscript to taped messages.
Local market   Loosely, having to do with the immediate market. More precisely as defined by the FCC, a market that generates 50% or more of its programs or commercials within its coverage area.
Log   Official record of a broadcast day, generally kept by automatic means, noting opening and closing times of all programs, commercials, and other nonprogram material.
Logo   Concise and striking image representing a station, network or system, or other media or nonmedia business.

Marketing   The collective term for advertising, promotion, direct mail, and direct sales.
Merchandizing   Selling or promoting by offering products, especially such giveaways as T-shirts, coffee mugs, and stickers, intended for audiences and client advertisers.
Mix session   (Also known as Mix-down) Production session where all the components of a promotional spot (or program or other announcement) are assembled; usually involves editing the video component and combining it with the audio portion; may also be a sound mix only, in which case it means assembling voices, music, and special effects.
MOR   Middle-of-the-Road radio format. (Also called variety, general appeal, diversified). The music it airs is not too current, not too old, not too loud, not too soft. It is all things to all people. Predominantly over-forty-five demographic.

Morning Drive   Radio's prime-time daypart: 6:00–10:00 A.M.
Multiple spot   An on-air promo containing mentions of two or more programs.
Music bed   Music track played under voice-over copy. See *bed*.
Music package   Specially produced music to be used in a promotional campaign.

NAB   National Association of Broadcasters: major trade association of the broadcast industry.
NBC   National Broadcasting Company; network formed in 1926 under the leadership of David Sarnoff.

Outdoor   Billboards, transit ads, signs, and display cards, especially on or in taxis, subways, trains, and buses, used for advertising or promotion.

Passives   Listeners who do not call stations in response to contests or promotions or to make requests or comments.
Penetration   Reach in a given population. The percentage of households using a product or receiving a service.
Playlist   The listing of all the songs to be played on a radio station's schedule. Today, it is computer-generated.
Plug   Promo, connector
Positioning   The programmer's attempt to place his or her station into a profitable and vacant niche in the area's radio offerings. Persuading the audience that one station or program network is really different from its competitors. Making your station second nature with your listeners.
Production   The use of studio equipment to combine sounds into a finished product.
Program director   (PD) Station executive usually in charge of all staff directly involved in putting material on the air. In some cases, a program director is limited to chief announcer duties.
Program guides   Magazines listing radio and television programs intended to aid listeners and viewers in choosing programs, and it also advertisers programs and images. (see *Guides*)
Promo   Spot announcement about a program, usually containing information about a contest, airtime, and program channel. May also give information or create a positive image for a personality or a station or network.
Promotion   Persuasive information about a program, personality station, or network using on-air and print media; intended to foster increased listening or viewing or a positive image.
Promotion (generic)   Promotions that focus on the overall qualities of a station (also known as general image promotion). Promotes the overall sound or image of a station.
Promotion kit   Folder or brochure, usually 9 × 12", prepared by a network, station, or syndicator to advertise one or more programs or a program schedule. (all the station's sports programs, for example). A typical "kit" will include program profiles, photos and biographies of stars, information on a show's pre-

vious ratings and audience appeal, and any thing else that is printable and convinces advertisers or program executives to close a sale.

Psychographics Information on the attitudes, interests, and opinions of audience members (also known as lifestyle data).

Publicity Public participation that brings visibility through an unpaid source. Unpaid mentions in newspapers, magazines, public speeches, or other stations.

Public Relations The process of building visibility and a positive image for a station.

RAB Radio Advertising Bureau, Inc. "The Radio Marketing Consultant's Guide to Media." New York, Radio Advertising Bureau, 1984.

Radio superstation FM broadcast station retransmitted by satellite or microwave to distant cable systems for local subscribers.

Rate card Current listing of a station's commercial time charges.

Rating Audience-measurement unit estimating the percentage of a total audience that tuned to a specific program or station during a 15-minute period (quarter-hour) or at least once during a week (cumulative rating).

Reach An indication of how many different members of an audience will be exposed to a message.

Remote Live production from locations other than a studio(such as football games, live news events, or broadcasts from a shopping mall).

Rep firm A national sales representative firm that deals with advertisers in large cities where it would be impractical for a local station to place a salesperson.

Royalty Compensation paid to copyright holder for the right to use copyrighted material.

Satellite Orbiting device for relaying audio from one earth station to another (e.g., DBS, Comsat, Satcom).

Secondary service area AM skywave listening area.

Signature Theme, logo, jingle, or ID.

SMP Strategic marketing plan: total marketing plan of a station involving sales, programming, promotion, public relations, advertising, etc.

Special One-time entertainment or news programs with special interest; usually applied to network programs that interrupt regular schedules.

Sponsor The advertiser featured in commercial announcements.

Spot A commercial announcement.

Station Facility operated by a licensee in order to broadcast radio or television signals on an assigned frequency.

Syndication The marketing of programs on a station-by-station basis (rather than through a network) to affiliates, independents, or pay television for a specific number.

Tag Short summary statement, usually a command to continue to tune in, placed at the end of a promo or program that tells the viewer or listener an upcoming program's title, day, and time, and the station's call letters.

Talent   Radio performer, announcer, DJ, newscaster.

Talk   Conversation and interview radio format.

Targeting   Aiming programs (generally by selecting appropriate appeals) at a demographically or psychographically defined audience.

Tease   Brief spot, usually ten seconds or less, intended to lure an audience into listening to or watching the succeeding program or news story by provoking interest. Placed near the end of the preceding program or story usually just before a commercial break. Also known as a *bumper*.

Tie-in package   Promo titles, music, graphics, and other materials reflecting a station or network's promotional campaign, sold as a full or partial package to an affiliate, permitting it to merge its local promotional look with that of its network.

Time-buyers   Advertising agency executives who purchase station time on behalf of their client.

Top 40   Radio music format consisting of continuous replay of the 40 highest-rated popular songs.

Topical promotion   Promotion tied to the content of the same day's or next day's episodes, guests (on talk shows), or news stories.

Trade-out   Exchange of station airtime for goods or services. Also called "recips" (reciprocal trades).

Traffic   Station department responsible for scheduling sponsor announcement.

Trafficking   The resale of a station within three years. FCC deregulation now makes it ineffective.

Upscale   People who rank toward the upper end of socioeconomic scales measuring income, education, and the like (work/profession, housing).

Vendor Plan   A cooperative sales promotion concept involving a client and the radio station.

Voice-over   Narrative audio either dubbed onto a videotape or added live while the tape plays. Frequently used for specific promotional tags.

Suggested Reading

Albert, James. *The Broadcaster's Legal Guide for Conducting Station Contests and Promotions*. Chicago: Bonus Books, 1986

Bergendorff, Fred L., Charles Harrison Smith and Lance Webster. *Broadcast Advertising and Promotion: A Handbook for Students and Professionals*. New York: Hastings House, 1983.

BPME *Big Ideas*. *BPME Newsletter* special reports. Each "Big Ideas" report covers a single subject of special interest to broadcast promoters with examples of supporting promotions from stations all over the United States.

Broadcast Financial Management Association and the National Association of Broadcasters. *1988 Radio Report*. Washington, D.C.: National Association of Broadcasters, 1988.

Eastman, Susan T., Sydney W. Head and Lewis Klein. *Broadcast/Cable Programming: Strategies and Practices*, 3rd ed. Belmont, CA: Wadsworth, 1989.

Eastman, Susan T. and Robert A. Klein. *Promotion and Marketing for Broadcast and Cable*, 2nd ed. Belmont, CA: Wadsworth, 1991.

Eastman, Susan T. and Robert A. Klein. *Strategies in Broadcasting Cable Promotion*. Belmont CA: Wadsworth, 1982.

Eicoff, Al. *Eicoff on Broadcast Direct Marketing*. Lincolnwood, IL: NTC Business Books, 1987.

Fornatale, Peter and Joshua E. Mills. *Radio in the Television Age*. Woodstock, NY: Overlook Press, 1980.

Fratello, Sharon and Cynthia Johnson. "Cashing in on Radio-Smart Research." *BPME IMAGE*, April/May 1989. Vol. 12.

Goff, Christine F., ed. *The Publicity Process*. Ames: Iowa State University Press, 1989.

Guidelines for Radio Copyrighting. Washington DC: National Association of Broadcasters, 1982.

Guidelines for Radio Promotion I. Washington, DC: National Association of Broadcasters, 1981.

Guidelines for Radio Promotion II. Washington, DC: National Association of Broadcasters, 1985.

Guidelines for Radio Promotion III. Washington DC: National Association of Broadcasters, 1988.

Hammerling, John. *It Only Takes One: How to Create the Right Idea*, New York: Simon & Schuster, 1983.

Helregel, Brenda. *Broadcasters' Public Service Activities*. Washington, DC: NAB Research and Planning Department, 1988.

Keith, Micheal C. and J.M. Krause. *The Radio Station*, 2nd ed. Boston: Focal Press, 1989.

Lotteries and Contests: A Broadcasters Handbook. Washington, DC: National Association of Broadcasters, 1990.

Marcus, Norman. *Broadcast and Cable Management*. Englewood Cliffs, NJ: Prentice-Hall, 1986.

Matelski, Marilyn J. *Broadcast Programming and Promotion Work-Text*. Stoneham, MA: Focal Press, 1989.

McCavitt, William E. and Peter K. Pringle. *Electronic Media Management*. Boston: Focal Press, 1986.

Mcdonald, Jack. *The Handbook of Radio Publicity and Promotion*, Blue Ridge Summit, TAB book, 1979.

McKinsey and Company. *Radio in Search of Excellence: Lessons From America's Best-run Radio Stations*. New York: Mckinsey and Company, 1985.

MacFarland, David T. *The Development of the Top 40 Format*. New York: Arno Press, 1979.

Meeske, Milan D. and R.C. Norris. *Copywriting for the Electronic Media: A Practical Guide*. Belmont CA: Wadsworth, 1987.

Moyes, William. *Successful Radio Promotions: From Ideas to Dollars*. Washington, DC: National Association of Broadcasters, 1988

NAB Legal Guide to Broadcast Law and Regulation. Washington, DC: National Association of Broadcasters, 1988.

O'Connell, Jim and Edye McIlhenney. "The Modern Promotion Director." *BPME IMAGE*, December 1988, 18.

"Radio '89: A Tale of Five Cities." *Broadcasting*, 11 September 1989, 75. Examines the marketing and promotion practices at five major-market radio stations.

Radio Only: The Monthly Management Tool. Cherry Hill, NJ: Inside Radio Inc., 1981.

Sherman, Barry. *Telecommunications Management*. New York: McGraw-Hill, 1987.

Sjodahl, Karl."Planning New Promotion Campaigns." *BPME Image*, August 1987, 31.

Sjodahl, Karl. "Station Logos: Building an Image." *Millimeter*, May 1982, 158.

Small Market Radio (newsletter), October 6, 1980, Otsego, MI.

Stone, Bob and John Wyman. *Successful Telemarketing*. Lincolnwood, IL: NTC Business Books, 1989.

Wells, William, John Burnett, and Sandra Moriarty. *Advertising: Principles and Practice*. Englewood Cliffs, NJ: Prentice-Hall, 1989.